Thomas William Allies

The Holy See and the Wandering of the Nations

From St. Leo to St. Gregory I

Thomas William Allies

The Holy See and the Wandering of the Nations
From St. Leo to St. Gregory I

ISBN/EAN: 9783337015091

Printed in Europe, USA, Canada, Australia, Japan

Cover: Foto ©ninafisch / pixelio.de

More available books at **www.hansebooks.com**

AND

THE WANDERING OF THE NATIONS

FROM ST. LEO I. TO ST. GREGORY I.

BY

THOMAS W. ALLIES, K.C.S.G.

AUTHOR OF THE "FORMATION OF CHRISTENDOM"; "CHURCH AND STATE AS SEEN IN THE FORMATION OF CHRISTENDOM"; "THE THRONE OF THE FISHERMAN"; "A LIFE'S DECISION"; AND "PER CRUCEM AD LUCEM"

LONDON: BURNS & OATES, Limited
NEW YORK: CATHOLIC PUBLICATION SOCIETY CO.
1888

THE LETTERS OF THE POPES AS SOURCES OF HISTORY.

CARDINAL MAI has left recorded his judgment that, "in matter of fact, the whole administration of the Church is learnt in the letters of the Popes".[1]

I draw from this judgment the inference that of all sources for the truths of history none are so precious, instructive, and authoritative as these authentic letters contemporaneous with the persons to whom they are addressed. The first which has been preserved to us is that of Pope St. Clement, the contemporary of St. Peter and St. Paul. It is directed to the Church of Corinth for the purpose of extinguishing a schism which had there broken out. In issuing his decision the Pope appeals to the Three Divine Persons to bear witness that the things which he has written "are written by us through the Holy Spirit," and claims obedience to them from those to whom he sends them as words "spoken by God through us".[2]

If the decisions of the succeeding Popes in the interval of nearly two hundred and fifty years between

---

[1] *Nova Patrum bibliotheca*, p. vi.: In Pontificum reapse epistolis tota ecclesiæ administratio cognoscitur.

[2] See p. 351 below; also *Church and State*, pp. 198-200, for the full statement of this passage.

this letter of St. Clement, about the year 95, and the great letter of St. Julius to the Eusebianising bishops at Antioch in 342, had been preserved entire, the constitution of the Church in that interval would have shone before us in clear light. In fact, we only possess a few fragments of some of these decisions, for there was a great destruction of such documents in the persecution which occupied the first decade of the fourth century. But from the time of Pope Siricius, in the reign of the great Theodosius, a continuous, though not a perfect, series of these letters stretches through the succeeding ages. There is no other such series of documents existing in the world. They throw light upon all matters and persons of which they treat. This is a light proceeding from one who lives in the midst of what he describes, who is at the centre of the greatest system of doctrine and discipline, and legislation grounded upon both, which the world has ever seen. One, also, who speaks not only with a great knowledge, but with an unequalled authority, which, in every case, is like that of no one else, but can even be *supreme*, when it is directed with such a purpose to the whole Church. Every Pope *can* speak, as St. Clement, the first of this series, speaks above, claiming obedience to his words as "words spoken by God through us".

In a former volume I made large use of the letters of Popes from Siricius to St. Leo. I have continued that use for the very important period from St. Leo to St. Gregory. Especially in treating of the Acacian schism I have gone to the letters of the Popes who had to deal with it—Simplicius, Felix III., Gelasius, Anastasius II., Symmachus, and Hormisdas. I have done the same for the important reign of Justinian; most of all for the grand pontificate of St. Gregory, which crowns the whole patristic period and sums up its discipline.

I am, therefore, indebted in this volume, first and chiefly, to the letters of the Popes and the letters addressed to them by emperors and bishops, stored up in Mansi's vast collection of Councils (1759, 31 volumes). I am also much indebted to Cardinal Hergenröther's work *Photius, sein Leben, und das griechische Schisma*, and to his *Handbuch der allgemeinen Kirchengeschichte*, as the number of quotations from him will show. Again, I may mention the two histories of the city of Rome, by Reumont and Gregorovius, as most valuable. I acknowledge many obligations to Riffel's *Geschichtliche Darstellung des Verhältnisses zwischen Kirche und Staat*, with regard to the legislation of Justinian. The edition of Justinian referred to by me is Heimbach's *Authenticum*, Leipsic, 1851. I have consulted Hefele's *Conciliengeschichte* where need was. I have found Kurth's *Origines de la Civilisation moderne* instructive. I have used the carefully emended and supplemented German edition of Röhrbacher's history, by various writers—Rump and others. St. Gregory is quoted from the Benedictine edition.

As these works are indicated in the notes as they occur with the single name of the author, I have given here their full titles.

The present volume is the sixth of the *Formation of*

Christendom, though it has a special title indicating the particular part of that general subject which it treats. I have, therefore, added to the numbering of the chapters in the Table of Contents the number which they hold in the whole work.

September 11, 1888.

TABLE OF CONTENTS.

CHAPTER I. (XLIII.).

The Holy See and the Wandering of the Nations.

	PAGE
Introduction. Connection with Volume V. St. Leo's action,	1
Denial of the Primacy as acknowledged at Chalcedon suicidal on the part of those who believe in the Church,	3
Subject of this volume as compared with the fifth,	5
The second wonder in human history,	6
The acknowledgment of the Primacy and the political powerlessness of the city of Rome coeval,	6
The three hundred years from Genseric to Astolphus,	9
St. Leo in Rome after Genseric,	10
Political condition of Rome. Avitus emperor, 455-6,	13
Majorian emperor, 457-461,	14
Death of Pope Leo; changes seen by him in his life,	15
Hilarus Pope and Libius Severus emperor, 461-465,	16
The over-lordship of Byzantium admitted in the choice of the Greek Anthemius as emperor, 467,	18
Sidonius Apollinaris an eye-witness of Rome's splendour, subjection to Byzantium, and unchanged habits in 467,	19
Anthemius murdered and Rome plundered by Ricimer, 472,	20
Olybrius emperor, 472; Ricimer and Olybrius die of the plague,	20
Glycerius emperor, 473; Nepos, 474; Romulus Augustulus, 475,	21

TABLE OF CONTENTS.

	PAGE
The senate declares to the eastern emperor that an emperor of the West is needless,	22
The twenty-one years' death-agony of imperial Rome,	23
State of the western provinces since the death of Theodosius I.,	24
The first and the second victory of the Church,	25
The effect produced by the wandering of the nations,	26
The Visigoth and Ostrogoth migrations,	27
Gaul overrun by Teuton invaders,	28
Arianism propagated by the Goths among the other tribes,	29
Burgundian kingdom of Lyons. Spain overrun,	30
The Vandals in North Africa and their persecution of Catholics,	31
The Hunnish inroads,	33
All the western provinces under Teuton governments,	35
Odoacer and Theodorick,	36
Odoacer succeeded by Theodorick after the capture of Ravenna,	38
The character of Theodorick's reign,	39
His fairness towards the Roman Church and Pontiff,	40
The contrast between Theodorick and Clovis,	42
The dictum of Ataulph on the Roman empire,	43
Ataulph and Theodorick represent the better judgments of the invaders,	44
The outlook of Pope Simplicius at Rome over the western provinces,	45
And over the eastern empire,	46
Basiliscus and Zeno the first theologising emperors,	47
How the races descending on the empire had become Arian,	49
The point of time when the Church was in danger of losing all which she had gained,	50
How the division of the empire called out the Primacy,	51
How the extinction of the western empire does so yet more,	53
How the Pope was the sole fixed point in a transitional world,	54
Guizot's testimony,	55
What St. Jerome, St. Augustine, and St. Leo did not foresee, which we behold,	57

CHAPTER II. (XLIV.)

Cæsar fell down.

	PAGE
Great changes in the Roman State following the time of St. Leo,	59
Nature of the succession in the Cæsarean throne, and then in the Byzantine,	61
Personal changes in the Popes and eastern emperors,	62
Gennadius succeeds Anatolius, and Acacius succeeds Gennadius in the see of Constantinople,	64
Acacius resists the Encyclikon of Basiliscus,	65
Letter of Pope Simplicius to the emperor Zeno,	66
Advancement of Acacius by Zeno,	69
Acacius induces Zeno to publish a formulary of doctrine,	70
John Talaia, elected patriarch of Alexandria, appeals for support to Pope Simplicius,	70
Pope Felix sends an embassy to the emperor,	71
His letter to Zeno,	72
His letter to Acacius,	73
His legates arrested, imprisoned, robbed, and seduced,	74
Pope Felix synodically deposes Acacius,	75
Enumerates his misdeeds in the sentence,	76
Synodal decrees in Italy signed by the Pope alone,	78
Letter of Pope Felix to Zeno setting forth the condemnation of Acacius,	79
The condition of the Pope when he thus wrote,	81
How Acacius received the Pope's condemnation,	83
The position which Acacius thereupon took up,	84
The greatness of the bishop of Constantinople identified with the greatness of his city,	84
The humiliations of Rome witnessed by Acacius,	86
How the Pope, under these humiliations, spoke to Acacius and to the emperor,	88
The Pope on the one side, Acacius on the other, represent an absolute contradiction,	89
Eudoxius and Valens matched by Acacius and Zeno,	92
Death of Acacius, and estimate of him by three contemporaries,	93

TABLE OF CONTENTS.

	PAGE
Fravita, succeeding Acacius, seeks the Pope's recognition,	93
Letters of the emperor and Fravita to the Pope, and his answers,	94
The position taken by Acacius not maintained by Zeno and Fravita,	96
Nor by Euphemius, who succeeds Fravita,	96
Euphemius suspects and resists the new emperor Anastasius,	97
Condition of the Empire and the Church at the accession of Pope Gelasius in 492,	98
The "libellus synodicus" on the emperor Anastasius,	100
With whom the four Popes—Gelasius, Anastasius, Symmachus, and Hormisdas—have to deal,	101
Euphemius, writing to the Pope, acknowledges him to be successor of St. Peter,	103
Gelasius replies to Euphemius, insisting on the repudiation of Acacius,	104
Absolute obedience of the Illyrian bishops professed to the Apostolic See,	105
Gelasius shows that the canons make the First See supreme judge of all,	106
Says that the bishop of Constantinople holds no rank among bishops,	107
Praises bishops who have resisted the wrongdoings of temporal rulers,	108
The Holy See, in virtue of its Principate, confirms every Council,	109
Gelasius in 494 defines to the emperor the domain of the Two Powers,	110
And the subordination of the temporal ruler in spiritual things,	111
The words of Gelasius have become the law of the Church,	113
The emperor Anastasius deposes Euphemius by the Resident Council,	114
Pope Gelasius, in a council of seventy bishops at Rome, sets forth the divine institution of the Primacy,	115
And the order of the three Patriarchal Sees,	115
And three General Councils—the Nicene, Ephesine, and Chalcedonic,	115

Denies to the see of Constantinople any rank beyond that
 of an ordinary bishop, and omits the Council of 381, . 116
Death of Pope Gelasius and character of his pontificate, . 118
His own description of the time in which he lived, . . 118

CHAPTER III. (XLV.).

Peter stood up.

Pope Anastasius : his letter to the emperor Anastasius, . 120
He makes the Pope's position in the Church parallel with
 that of the emperor in the world, ... 121
He writes to Clovis on his conversion, ... 122
St. Gregory of Tours notes the prosperity of Catholic king-
 doms and the decline of Arian in the West, . . . 123
Letter of St. Avitus, bishop of Vienne, to Clovis on his
 baptism, ... 124
He recognises the vast importance of the professing the
 Catholic faith by Clovis, ... 125
And the duty of Clovis to propagate the faith in peoples
 around, ... 126
How the words of St. Avitus to Clovis were fulfilled in
 history, ... 127
The election of Pope Symmachus traversed by the emperor's
 agent, ... 128
His letter termed "Apologetica" to the eastern emperor, . 129
The imperial and papal power compared, ... 131
The papal and the sovereign power the double permanent
 head of human society, ... 133
Emperors wont to acknowledge Popes on their accession, . 134
Inferences to be deduced from this letter, ... 135
The answer of the emperor Anastasius is to stir up a fresh
 schism at Rome, ... 136
The Synodus Palmaris, without judging the Pope, declares
 him free from all charge, ... 137
Letter of the bishop of Vienne to the Roman senate upon
 this Council, ... 139

	PAGE
The cause of the Bishop of Rome is not that of one bishop, but of the Episcopate itself,	140
Words of Ennodius, bishop of Pavia, embodied in the acts of the Roman Council of 503,	142
Result of the attack of the emperor on the Pope is the recording in black and white that the First See is judged by no man,	143
The eastern Church under the emperor Anastasius,	143
He deposes Macedonius as well as Euphemius,	144
Both these bishops of Byzantium failed to resist his despotism,	147
Eastern bishops address Pope Symmachus to succour them,	148
Pope Hormisdas succeeds Symmachus in 514,	149
His instruction to the legates sent to Constantinople,	150
The bishop of Constantinople presents all bishops to the emperor,	157
The conditions for reunion made by Pope Hormisdas,	158
The treacherous conduct of the emperor,	159
Hormisdas describes Greek diplomacy,	160
The Syrian Archimandrites supplicate the Pope for help,	161
Sudden death of the emperor Anastasius,	162
The emperor Justin's election and antecedents,	162
He notifies his accession to the Pope,	163
The Pope holds a council and sends an embassy to Constantinople,	164
The bishop, clergy, and emperor accept the terms of the Pope,	165
The formulary of union signed by them,	167
The report of the legates to the Pope,	169
The emperor Justin's letter to the Pope,	170
Character of the period 455-519,	171
Political state of the East and West most perilous to the Church,	172
The Popes under Odoacer and Theodorick,	173
How Acacius took advantage of the political situation,	174
The meaning and range of his attempt,	175
The Pope from 476 onwards rests solely upon his Apostolate,	176
The seven Popes who succeed St. Leo,	179

	PAGE
The seven bishops who succeed Anatolius at Constantinople,	180
The eastern emperors in this time,	182
The state of the eastern patriarchates, Alexandria and Antioch,	184
The waning of secular Rome reveals the power of the Pontificate,	185
The Popes alone preserved the East from the Eutychean heresy,	185
The position of St. Leo maintained by the seven following Popes,	186
The submission to Hormisdas an act of the "undivided" Church,	187
The adverse circumstances which developed the Pope's Principate,	188

CHAPTER IV. (XLVI.).

JUSTINIAN.

Sequel in Justinian of the submission to Pope Hormisdas,	189
His acknowledgment of the Primacy to Pope John II. in 533,	190
Reply of Pope John II. confirming the confession sent to him by Justinian,	191
The *Pandects* of Justinian issued in the same year,	192
Close interweaving of ecclesiastical and temporal interests,	193
Interference with the freedom of the papal election by the temporal ruler,	194
Letter of Cassiodorus as Prætorian prefect to Pope John II.,	195
Justinian all his reign acknowledged the Primacy of the Pope,	196
His character, purposes, and actions,	196
Succeeds his uncle the emperor Justin I.,	198
Great political changes coeval with his succession,	199
He reconquers Northern Africa by Belisarius,	199
The Catholic bishops of Africa meet again in General Council,	200

	PAGE
They send an embassy to consult Pope John II.,	201
Pope Agapetus notes their reference to the Apostolic Principate,	202
Great renown of Justinian at the reconquest of Africa,	203
Pope Agapetus at Constantinople deposes its bishop,	204
Justinian begins the Gothic War. Belisarius enters Rome,	205
He is welcomed as restorer of the empire,	206
The empress Theodora deposes Pope Silverius by Belisarius,	207
First siege of Rome by Vitiges,	210
The mausoleum of Hadrian stripped of its statues,	211
Vitiges, having lost half his army, raises the siege,	213
Belisarius, having reconquered Italy, is recalled for the war with Persia,	214
Totila, elected Gothic king, renews the war,	214
Visits St. Benedict at Monte Cassino, and is warned by him,	215
Second siege of Rome by Totila,	216
Rome taken by Totila in 546,	216
Third capture of Rome by Belisarius, in 547,	217
Fourth capture of Rome by Totila, in 549,	218
Totila defeated and killed by Narses at Taginas,	219
Fifth capture of Rome by Narses, in 552,	220
End of the Gothic war, in 555,	221
Its effect on the civil condition of the Pope, Italy, and Rome,	222
The sufferings of Rome from assailants and defenders,	223
The new test of papal authority applied by these events,	225
Vigilius, having become legitimate Pope, is sent for by Justinian,	226
Church proceedings at Constantinople after the death of Pope Agapetus,	227
The patriarch Mennas, in conjunction with the emperor, consecrates at Constantinople a patriarch of Alexandria,	228
The Origenistic struggle in the eastern empire,	229
Justinian theologising,	230
The whole East urged to consent to his edict on doctrine,	231
Pope Vigilius, summoned by Justinian, enters Constantinople,	232
After long conferences with emperor and bishops he issues a Judgment,	234

The Pope and emperor agree upon holding a General Council,	235
The emperor's despotism, and the bishops crouching before it,	236
The Pope takes sanctuary, and is torn away from the altar,	237
Flies to the church at Chalcedon,	238
The bishops relent, and the Pope returns to Constantinople,	239
Eutychius, succeeding Mennas, proposes a council under presidency of the Pope,	239
The emperor causes it to meet under Eutychius without the Pope,	240
Proceedings of the Council. The Pope declines their invitation,	241
Close of the Council, without the Pope's presence,	242
The Pope issues a Constitution apart from the Council,	242
Also a condemnation of the Three Chapters without mention of the Council,	243
The Pope on his way back to Rome dies at Syracuse,	244
The patriarch Eutychius, refusing to sign a doctrinal decree of Justinian, is deposed by the Resident Council,	244
Justinian issues his Pragmatic Sanction for government of Italy,	245
State of things following in Italy,	246
Justinian's conception of the relation between Church and State,	248
He gives to the decrees of Councils and to the canons the force of law,	250
Three leading principles in these enactments,	251
The State completely recognises the Church's whole constitution,	251
The episcopal idea thoroughly realised,	253
Concurrent action of the laws of Church and State herein,	254
Justinian further associated bishops with the civil government,	255
The part given to them in civil administration,	256
A system of mutual supervision in bishops and governors,	257
The branches of civil matters specially put under bishops,	259
The completeness and the cordiality of the alliance with the Church,	261

	PAGE
Which differentiates Justinian's attitude from that of modern governments,	262
In what Justinian was a true maintainer of the imperial idea,	264
The dark blot which lies upon Justinian,	267
How he passed from the line of defence to that of interference and mastery,	269
The result, spiritual and temporal, of Justinian's reign,	270

CHAPTER V. (XLVII.).

St. Gregory the Great.

The state of Rome as a city after the prefecture of Narses,	272
Contrast of Nova Roma,	274
The Rome of the Church a new city,	275
St. Gregory's antecedents as prefect, monk, nuncio, and deacon of the Roman Church,	276
Elected Pope against his will. His description of his work,	278
And of the time's calamity,	279
The utter misery of Rome expressed in the words of Ezechiel,	281
Contrast between the language used of Rome by St. Leo and St. Gregory,	283
St. Gregory closes his preaching in St. Peter's, overcome with sorrow,	284
The works of St. Gregory out of this Rome,	285
The Lombard descent on Italy,	287
Rome ransomed from the Lombards, and Monte Cassino destroyed,	290
The Primacy untouched by the temporal calamities of Rome,	292
Its unique prerogative brought out by unequalled sufferings,	293
The new city of Rome lived only by the Primacy,	294
St. Gregory's account of the Primacy to the empress Constantina,	295
He identifies his own authority with that of St. Peter,	296
Writes to the emperor Mauritius that the union of the Two Powers would secure the empire against barbarians,	297

	PAGE
Claims to the emperor St. Peter's charge over the whole Church,	298
John the Faster's assumed title an injury to the whole Church,	299
What St. Gregory infers from the three patriarchal sees being all sees of Peter,	301
Contrast drawn by St. Gregory between the Pope's Principate and John the Faster's assumed title,	302
The fatal falsehood which this title presupposed,	303
The opposing truth in the Principate made *de Fide* by the Vatican Council,	306
St. Leo against Anatolius, and St. Gregory against John the Faster, occupy like positions,	307
St. Gregory's title, "Servant of the servants of God," expresses the maxim of his government,	308
The fourteen books of St. Gregory's letters range over every subject in the whole Church,	309
The special relation between the sees of St. Peter and St. Mark,	311
Asserts his supremacy to the Lombard queen Theodelinda,	311
St. Gregory appoints the bishop of Arles to be over the metropolitans of Gaul,	312
The venture of St. Gregory in attempting the conversion of England,	313
St. Augustine commended to queen Brunechild and consecrated by the bishop of Arles, and the English Church made by Gregory,	315
Work of St. Gregory in the Spanish Church,	316
He relates the martyrdom of St. Hermenegild,	316
His letters to St. Leander of Seville,	317
Conversion of king Rechared,	318
St. Gregory's letter of congratulation to him,	318
Letter of king Rechared informing the Pope of his conversion,	321
Gibbon's account of the government which was the result of Rechared's conversion,	322
The important principles thus consecrated by the Church,	324

TABLE OF CONTENTS.

	PAGE
Overthrow of the Arian kingdoms in Africa, Spain, Gaul, and Italy, between Pope Felix III. and Pope Gregory I.,	325
The equal failure of Genseric, Euric, Gondeband, and Theodorick,	327
The part in this which the Catholic bishops had,	329
The Spanish monarchy first of many formed by the Church,	331
Superiority of this government to the Byzantine absolutism,	332
St. Gregory as fourth doctor of the western Church,	334
St. Gregory as a chief artificer in the Church's second victory,	335
Summary of St. Gregory's action as metropolitan patriarch and Pope,	337
Councils held by him in Rome: protection of monks,	338
His management of the Patrimonium Petri,	340
His success with schismatics and heretics,	341
The Primacy from St. Leo to St. Gregory,	342
The continued rise of the bishop of Constantinople,	343-5
The political degradation and danger of Rome,	345
Long disaster reveals still more the purely spiritual foundation of the Primacy,	346
Testimony given by the disappearance of the Arian governments and the conversion of Franks and Saxons,	347
The patriarchate of Constantinople imposed by civil law,	348
The Nicene constitution in the East impaired by despotism and heresy,	349
The persistent defence of this constitution by the Popes,	350
The Petra Apostolica in the sixty Popes preceding Gregory,	352
As discerned by Hurter in the time of Pope Innocent III.,	353
As in the time from Pope Innocent III. to Leo XIII.,	355
The continuous Primacy from St. Peter to St. Gregory,	355
As Rome diminishes the Primacy advances,	356
The times in which it was exercised by St. Gregory,	358
The opposing forces which unite to sustain the Petra Apostolica,	359
INDEX,	361

# THE HOLY SEE AND THE WANDERING OF THE NATIONS.

CHAPTER I.

THE HOLY SEE AND THE WANDERING OF THE NATIONS.

> "Rome's ending seemed the ending of a world.
> If this our earth had in the flat sea sunk,
> Save one black ridge whereon I sat alone,
> Such wreck had seemed not greater. It was gone,
> That empire last, sole heir of all the empires,
> Their arms, their arts, their letters, and their laws.
> The fountains of the nether deep are burst,
> The second deluge comes. And let it come!
> The God who sits above the waterspouts
> Remains unshaken."
>
> —A. DE VERE, *Legends and Records*—"Death of St. Jerome".

I ENDED the last chapter by drawing out that series of events in the Church's internal constitution and of changes in the external world of action outside and independent of the Church which combined in one result the exhibition to all and the public acknowledgment by the Church of the Primacy given by our Lord to St. Peter, and continued to his successors in the See of Rome. I showed St. Leo as exercising this Primacy by annulling the acts of an Ecumenical

Council, the second of Ephesus, legitimately called and attended by his own legates, because it had denied a tenet of what St. Leo declared in a letter sent to the bishops and accepted by them to be the Christian faith upon the Incarnation itself. I showed him supported by the Church in that annulment, by the eastern episcopate, which attended the Council of Chalcedon, and by the eastern emperor, Marcian. Again, I showed him confirming the doctrinal decrees of the Ecumenical Council of Chalcedon, which followed the Council annulled by him, while he reversed and disallowed certain canons which had been irregularly passed. This he did because they were injurious to that constitution of the Church which had come down from the Apostles to his own time. And this act of his, also, I showed to be accepted by the bishop of Constantinople, who was specially affected, and by the eastern emperor, and by the episcopate: and also that the confirmation of doctrine on the one hand, and the rejection of canons on the other, were equally accepted. I also showed this great Council in its Synodical Letter to the Pope acknowledging spontaneously that very position of the Pope which the Popes had always set forth as the ground of all the authority which they claimed. The Council of Chalcedon addressed St. Leo "as entrusted by the Saviour with the guardianship of the Vine". But the Vine in the universal language of the Fathers betokened the whole Church of God. And the Council refers the confirmation of its acts to the Pope in the same document in which it asserts that the guardianship of the Vine was given to him by the Saviour Himself. This expression, "by the Saviour Himself," means that it was not given to him by the decree of any Council representing the Church. It is a full acknowledgment that the promises made to Peter, and the Pastorship conferred upon him, descended to his successor in the See of Rome. It is a full acknowledgment; for how else was St. Leo entrusted by the Saviour with the guardianship of the Vine? Those who so addressed him were equally bishops with himself; they equally enjoyed the one indivisible episcopate, "of which a part is held by each without division of the whole".[1] But this one, beside and beyond that, was charged with the whole—the Vine itself. This one point is that in which St. Peter went beyond his brethren, by the special gift and appointment of the Saviour Himself. The words, then, of the Council contain a special acknowledgment that the line of Popes after a succession of four hundred years sat in the person of Leo on the seat of St. Peter, with St. Peter's one sovereign prerogative.

It is requisite, I think, distinctly to point out that Christians, whoever they are, provided only that they admit, as confessing belief in any one of the three creeds, the Apostolic, the Nicene, or the Athanasian, they do admit, that there is one holy Catholic Church, commit a suicidal act in denying the Primacy as acknowledged by the Church at the Council of Chal-

---

[1] "Episcopatus unus est cujus a singulis in solidum pars tenetur."—S. Cyprian, *De Unitate Ecclesiæ*.

cedon. For such a denial destroys the authority of the Church herself both in doctrine and discipline for all subsequent time. If the Church, in declaring St. Leo to be entrusted by our Lord with the guardianship of the Vine, erred; if she asserted a falsehood, or if she favoured an usurpation, how can she be trusted for any maintenance of doctrine, for any administration of sacraments, for any exercise of authority? This consideration does not touch those who believe in no Church at all. They are in the position of that individual whom the great Constantine recommended to take a ladder and mount to heaven by himself. But it touches all who profess to believe in an episcopate, in councils, in sacraments, in an organised Church, in authority deposited in that Church, and, finally, in history and in historical Christianity. To all such it may surely be said, as the simplest enunciation of reasoning, that they cannot profess belief in the Church which the Creed proclaims while they accept or reject its authority as they please. Or to localise a general expression: A man does not follow the doctrine of St. Augustine if he accepts his condemnation of Pelagius, but denies that unity of the Church in maintaining which St. Augustine spent his forty years of teaching. The action of all such persons in the eyes of the world without amounts to this, that by denying the Primacy they disprove the existence of the Church. Their negation goes to the profit of total unbelief. Asserters of the Church's division are pioneers of infidelity, for who can believe in what has fallen? or is the kingdom of our Lord Jesus Christ a kingdom divided against itself? They who maintain schism generate agnostics.

But I was prevented on a former occasion by want of space from dwelling with due force upon some circumstances of St. Leo's life. These are such as to make his time an era. I was occupied during a whole volume with the attempt to set forth in some sort the action of St. Peter's See upon the Greek and Roman world from the day of Pentecost to the complete recognition of the Universal Pastorship of Peter as inherited by the Roman Pontiff in the person of St. Leo.

I approach now a further development of this subject. I go forward to treat of the Papacy, deprived of all temporal support from the fall of the western empire, taking up the secular capital into a new spiritual Rome, and creating a Christendom out of the northern tribes who had subverted the Roman empire.

There is, I think, no greater wonder in human history than the creation of a hierarchy out of the principle of headship and subordination contained in our Lord's charge to Peter. It has been pointed out that the constitution of the Nicene Council itself manifested this principle, and was the proof of its spontaneous action in the preceding centuries, while its overt recognition, as seated in the Roman Pontiff, is seen in the pontificate of St. Leo.

There is a second wonder in human history, on which it is the purpose of this volume to dwell. The Roman empire, in which the Pax Romana had provided a mould of widespread civilisation for the Church's growth, was at length broken up in the western half of it, by Teuton invaders occupying its provinces. These were all, at the time of their settlement, either pagan or Arian. There followed, in a certain lapse of time, the creation of a body of States whose centre of union and belief was the See of Peter. That is the creation of Christendom proper. The wonder seen is that the northern tribes, impinging on the empire, and settling on its various provinces like vultures, became the matter into which the Holy See, guiding and unifying the episcopate, maintaining the original principle of celibacy, and planting it in the institute of the religious life through various countries depopulated or barbarous, infused into the whole mass one spirit, so that Arians became Catholics, Teuton raiders issued into Christian kings, savage tribes thrown upon captive provincials coalesced into nations, while all were raised together into, not a restored empire of Augustus, but an empire holy as well as Roman, whose chief was the Church's defender *(advocatus ecclesiæ)*, whose creator was the Roman Peter.

It is not a little remarkable that this signal recognition by the Fourth General Council of the Roman Pontiff's authority coincided in time with the utter powerlessness to which Rome as a city was reduced. That city, on whose glory as queen of nations and civiliser of the earth her own bishop had dwelt with all the fondness of a Roman, when, year by year, on the feast of St. Peter and St. Paul, he addressed the assembled episcopate of Italy, ran twice, in his own time, the most imminent danger of ceasing to exist.

Italy was absolutely without an army to give her strongest cities a chance of resisting the desolation of Attila. Rome was without a force raised to save it from the pitiless robbery of Genseric. Without escort, and defended only by his spiritual character, Leo went forth to appeal before Attila for mercy to a heathen Mongol. There is no record of what passed at that interview. Only the result is known. The conqueror, who had swept with remorseless cruelty the whole country from the Euxine to the Adriatic Sea, who was now bent upon the seizure of Italy itself, and in his course had just destroyed Aquileia, was at Mantua marching upon Rome. His intention was proclaimed to crown all his acts of destruction with that of Rome. This was the dowry which he proposed to take for the hand of the last great emperor's granddaughter, proffered to him by the hapless Honoria herself. At the word of Leo the Scourge of God gave up his prey: he turned back from Italy, and relinquished Rome, and Leo returned to his seat. In the course of the next three years he confirmed, at the eastern emperor's repeated request, the doctrinal decrees of the great Council; but he humbled likewise the arrogance of Anatolius, and not all the loyalty of Marcian, not all the devotion of the empress and saint Pulcheria, could induce him to exalt the bishop of the eastern capital at the expense of the Petrine hierarchy. But during those same three years he saw, in Rome itself, Honoria's brother, the grandson of Theodosius, destroy his own throne, and thereupon the murderer of an emperor compel his widow to accept him in her husband's place, in the first days of her sorrow. He saw, further, that daughter of Theodosius and Eudoxia, when she learnt that the usurper of her husband's throne was likewise his murderer, call in the Vandal from Carthage to avenge her double dishonour. This was the Rome which awaited, trembling and undefended, the most profligate of armies, led by the most cruel of persecutors. Once more St. Leo, stripped of all human aid, went forth with his clergy on the road to the port by which Genseric was advancing, to plead before an Arian pirate for the preservation of the capital of the Catholic faith. He saved his people from massacre and his city from burning, but not the houses from plunder. For fourteen days Rome was subject to every spoliation which African avarice could inflict. Again, no record of that misery has been kept; but the hand of Genseric was heavier than that of Alaric, in proportion as the Vandal was cruel where the Ostrogoth was generous. Alaric would have fought for Rome as Stilicho fought, had he continued to be commanded by that Theodosius who made him a Roman general; but Genseric was the vilest in soul of all the Teuton invaders, and for fifty years, during the utter prostration of Roman power, he infested all the shores of the Mediterranean with the savagery afterwards shown by Saracen and Algerine.

This second plundering of Rome was no isolated event. It was only the sign of that utter impotence into which Roman power in the West had fallen. The city of Rome was the trophy of Cæsarean government during five hundred years—from Julius, the most royal, to Valentinian, the most abject of emperors. And now its temporal greatness was lost for ever. It ceased to be the imperial city, but by the same stroke became from the secular a spiritual capital. The Pope, freed from the western Cæsar,[1] gave to the Cæsarean city its second and greater life: a life of another kind generating also an empire of another sort. The raid of Genseric in the year 455 is the first of three hundred years of warfare carried on from the time of the Vandal through the time of the Lombard, under the neglect and oppression of the Byzantine, until, in the year 755, Astolphus, the last, and perhaps the worst, of an evil brood, laid waste the campagna, and besieged the city. St. Leo, in his double embassy to Attila and Genseric, was an unconscious prophet of the time to come, a visible picture of three hundred years as singular in their conflict and their issue as those other three hundred which had their close in the Nicene Council. During all those ages the Pope is never secure in his own city. He sees the trophy of Cæsarean empire slowly perish away. The capital of the world ceases to be even the capital of a province. The eastern emperor, who still called himself emperor of the Romans, omitted for many generations even to visit the city which he had subjected to an impotent but malignant official, termed an Exarch, who guarded himself by

---

[1] Gregorovius, i. 286. "Das Papstthum, vom Kaiser des Abendlandes befreit, erstand, und die Kirche Roms wuchs unter Trümmern mächtig empor. Sie trat an die Stelle des Reichs."

the marshes of Ravenna, but left Rome to the inroads of the Lombards. The last emperor who deigned to visit the old capital of his empire came to it only to tear from it the last relic of imperial magnificence. But then Jerusalem had fallen into the hands of the infidel, and Christian pilgrims, since they could no longer visit the sepulchre of Christ, flocked to the sepulchre of his Vicar the Fisherman. And thus Rome was become the place of pilgrimage for all the West. Saxon kings and queens laid down their crowns before St. Peter's threshold, invested themselves with the cowl, and died, healed and happy, under the shadow of the chief Apostle. When the three hundred years were ended, the arm of Pepin made the Pope a sovereign in his own newly-created Rome. During these three centuries, running from St. Leo meeting Genseric, the pilot of St. Peter's ship has been tossed without intermission on the waves of a heaving ocean, but he has saved his vessel and the freight which it bears—the Christian faith. And in doing this he has made the new-created city, which had become the place of pilgrimage, to be also the centre of a new world.

As Leo came back from the gate leading to the harbour and re-entered his Lateran palace, undefended Rome was taken possession of by the Vandal. Leo for fourteen days was condemned to hear the cries of his people, and the tale of unnumbered insults and iniquities committed in the palaces and houses of Rome. When the stipulated days were over, the plunderer bore away the captive empress and her daughters from the palace of the Cæsars, which he had so completely sacked that even the copper vessels were carried off. Genseric also assaulted the yet untouched temple of Jupiter on the Capitol, and not only carried away the still remaining statues in his fleet which occupied the Tiber, but stripped off half the roof of the temple and its tiles of gilded bronze. He took away also the spoils of the temple at Jerusalem, which Vespasian had deposited in his temple of peace. Belisarius found them at Carthage eighty years later, and sent them as prizes to Constantinople.[1]

Many thousand Romans of every age and condition Genseric carried as slaves to Carthage, together with Eudocia and her daughters, the eldest of whom Genseric compelled to marry his son Hunnerich. After sixteen years of unwilling marriage Eudocia at last escaped, and through great perils reached Jerusalem, where she died and was buried beside her grandmother, that other Eudocia, the beautiful Athenais whom St. Pulcheria gave to her brother for bride, and whose romantic exaltation to the throne of the East ended in banishment at Jerusalem. But one of the great churches at Rome is connected with her memory: since the first Eudocia sent to the empress her daughter at Rome half of the chains which had bound St. Peter at his imprisonment by Agrippa. When Pope Leo held the relics, which had come from Jerusalem, to those other relics belonging to the Apostle's captivity at Rome on his martyrdom, they grew together and

[1] Gregorovius, i. 200.

became one chain of thirty-eight links. Upon this the empress in the days of her happiness built the Church of St. Peter ad Vincula to receive so touching a memorial of the Apostle who escaped martyrdom at Jerusalem to find it at Rome. Upon his delivery by the angel "from all the expectation of the people of the Jews," he "went to another place". There, to use the words of his own personal friend and second successor at Antioch, he founded "the church presiding over charity in the place of the country of the Romans,"[1] and there he was to find his own resting-place. The church was built to guard the emblems of the two captivities. The heathen festival of Augustus, which used to be kept on the 1st August at the spot where the church was founded, became for all Christendom the feast of St. Peter's Chains.[2]

In the life of St. Leo by Anastasius, we read that after the Vandal ruin he supplied the parish churches of Rome with silver plate from the six silver vessels, weighing each a hundred pounds, which Constantine had given to the basilicas of the Lateran, of St. Peter, and of St. Paul, two to each. These churches were spared the plundering to which every other building

---

[1] St. Ignatius, *Epistle to the Romans.*
[2] "That Roman, that Judean bond
  United then dispart no more—
  Pierce through the veil; the rind beyond
  Lies hid the legend's deeper lore.
  Therein the mystery lies expressed
  Of power transferred, yet ever one;
  Of Rome—the Salem of the West—
  Of Sion, built o'er Babylon."
  A. de Vere, *Legends and Records*, p. 204.

was subjected. But the buildings of Rome were not burnt, though even senatorian families were reduced to beggary, and the population was diminished through misery and flight, besides those who were carried off to slavery.

At this point of time the grandeur of Trajan's city [1] began to pass into the silence and desolation which St. Gregory in after years mourned over in the words of Jeremias on ruined Jerusalem.

Let us go back with Leo to his patriarchal palace, and realise if we can the condition of things in which he dwelt at home, as well as the condition throughout all the West of the Church which his courage had saved from heresy.

The male line of Theodosius had ended with the murder of Valentinian in the Campus Martius, March 16, 455. Maximus seized his throne and his widow, and was murdered in the streets of Rome in June, 455, at the end of seventy-seven days. When Genseric had carried off his spoil, the throne of the western empire, no longer claimed by anyone of the imperial race, became a prey to ambitious generals. The first tenant of that throne was Avitus, a nobleman from Gaul, named by the influence of the Visigothic king, Theodorich of Toulouse. He assumed the purple at Arles, on the 10th July, 455. The Roman senate, which clung to its hereditary right to name the princes, accepted him, not being able to help itself, on the 1st January, 456; his son-in-law, Sidonius Apollinaris, delivered the

[1] Gregorovius, i. 208.

customary panegyric, and was rewarded with a bronze statue in the forum of Trajan, which we thus know to have escaped injury from the raid of Genseric. But at the bidding of Ricimer, who had become the most powerful general, the senate deposed Avitus; he fled to his country Auvergne, and was killed on the way in September, 456.

All power now lay in the hands of Ricimer. He was by his father a Sueve; by his mother, grandson of Wallia, the Visigothic king at Toulouse. With him began that domination of foreign soldiery which in twenty years destroyed the western empire. Through his favour the senator Majorian was named emperor in the spring of 457. The senate, the people, the army, and the eastern emperor, Leo I., were united in hailing his election. He is described as recalling by his many virtues the best Roman emperors. In his letter to the senate, which he drew up after his election in Ravenna, men thought they heard the voice of Trajan. An emperor who proposed to rule according to the laws and tradition of the old time filled Rome with joy. All his edicts compelled the people to admire his wisdom and goodness. One of these most strictly forbade the employment of the materials from older buildings, an unhappy custom which had already begun, for, says the special historian of the city, the time had already come when Rome, destroying itself, was made use of as a great chalk-pit and marble quarry;[1] and for such it served the Romans themselves for more

[1] Gregorovius, i. 215.

than a thousand years. They were the true barbarians who destroyed their city.

But Majorian was unable to prevent the ruin either of city or of state. He had made great exertions to punish Genseric by reconquering Africa. They were not successful; Ricimer compelled him to resign on the 2nd of August, 461, and five days afterwards he died by a death of which is only known that it was violent. A man, says Procopius, upright to his subjects, terrible to his enemies, who surpassed in every virtue all those who before him had reigned over the Romans.

Three months after Majorian, died Pope St. Leo. First of his line to bear the name of Great, who twice saved his city, and once, by the express avowal of a successor, the Church herself, Leo carried his crown of thorns one-and-twenty years, and has left no plaint to posterity of the calamities witnessed by him in that long pontificate. Majorian was the fourth sovereign whom in six years and a half he had seen to perish by violence. A man with so keen an intellectual vision, so wise a measure of men and things, must have fathomed to its full extent the depth of moral corruption in the midst of which the Church he presided over fought for existence. This among his own people. But who likewise can have felt, as he did, the overmastering flood of northern tribes—*vis consili expers*—which had descended on the empire in his own lifetime. As a boy he must have known the great Theodosius ruling by force of mind that warlike but savage host of Teuton mercaries. In his one life, Visigoth and

Ostrogoth, Vandal and Herule, Frank and Aleman, Burgundian and Sueve, instead of serving Rome as soldiers in the hand of one greater than themselves, had become masters of a perishing world's mistress; and the successor of Peter was no longer safe in the Roman palace which the first of Christian emperors had bestowed upon the Church's chief bishop. Instead of Constantine and Theodosius, Leo had witnessed Arcadius and Honorius; instead of emperors the ablest men of their day, who could be twelve hours in the saddle at need, emperors who fed chickens or listened to the counsel of eunuchs in their palace. Even this was not enough. He had seen Stilicho and Aetius in turn support their feeble sovereigns, and in turn assassinated for that support; and the depth of all ignominy in a Valentinian closing the twelve hundred years of Rome with the crime of a dastard, followed by Genseric, who was again to be overtopped by Ricimer, while world and Church barely escape from Attila's uncouth savagery. But Leo in his letters written in the midst of such calamities, in his sermons spoken from St. Peter's chair, speaks as if he were addressing a prostrate world with the inward vision of a seer to whom the triumph of the heavenly Jerusalem is clearly revealed, while he proclaims the work of the City of God on earth with equal assurance.

Hilarus in that same November, 461, succeeded to the apostolic chair. Hilarus was that undaunted Roman deacon and legate who with difficulty saved his life at the Robber-Council of Ephesus, where St.

Flavian, bishop of Constantinople, was beaten to death by the party of Dioscorus, and who carried to St. Leo a faithful report of that Council's acts. At the same time the Lucanian Libius Severus succeeded to the throne. All that is known of him is that he was an inglorious creature of Ricimer, and prolonged a government without record until the autumn of 465, when his maker got tired of him. He disappeared, and Ricimer ruled alone for nearly two years. Yet he did not venture to end the empire with a stroke of violence, or change the title of Patricius, bestowed upon him by the eastern emperor, for that of king. In this death-struggle of the realm the senate showed courage. The Roman fathers in their corporate capacity served as a last bond of the State as it was falling to pieces; and Sidonius Apollinaris said of them that they might rank as princes with the bearer of the purple, only, he adds significantly, if we put out of question the armed force.[1] The protection of the eastern emperor, Leo I., helped them in this resistance to Ricimer. The national party in Rome itself called on the Greek emperor for support. The utter dissolution of the western empire, when German tribes, Burgundians, Franks, Visigoths, and Vandals, had taken permanent possession of its provinces outside of Italy, while the violated dignity of Rome sank daily into greater impotence, now made Byzantium come forth as the true head of the empire. The better among the eastern Cæsars

---

[1] Sidonius Apollinaris, *Epist.*, i. 9. "Hi in amplissimo ordine, seposita prærogativa partis armatæ, facile post purpuratum principem principes erant."

acknowledged the duty of maintaining it one and indivisible. They treated sinking Italy as one of their provinces, and prevented the Germans from asserting lordship over it.

At length, after more than a year's vacancy of the throne, Ricimer was obliged not only to let the senate treat with the Eastern emperor, Leo I., but to accept from Leo the choice of a Greek. Anthemius, one of the chief senators at Byzantium, who had married the late emperor Marcian's daughter, was sent with solemn pomp to Rome, and on the 12th April, 467, he accepted the imperial dignity in the presence of senate, people, and army, three miles outside the gates. Ricimer also condescended to accept his daughter as his bride, and we have an account of the wedding from that same Sidonius Apollinaris who a few years before had delivered the panegyric upon the accession of his own father-in-law, Avitus, afterwards deposed and killed by Ricimer; moreover, he had in the same way welcomed the accession of the noble Majorian, destroyed by the same Ricimer. Now on this third occasion Sidonius describes the whole city as swimming in a sea of joy. Bridal songs with fescennine licence resounded in the theatres, market-places, courts, and gymnasia. All business was suspended. Even then Rome impressed the Gallic courtier-poet with the appearance of the world's capital. What is important is that we find this testimony of an eye-witness, given incidentally in his correspondence, that Rome in her buildings was still in all her splendour. And again in his long panegyric he makes Rome address the eastern emperor, beseeching him, in requital for all those eastern provinces which she has given to Byzantium—"Only grant me Anthemius;[1] reign long, O Leo, in your own parts, but grant me my desire to govern mine". Thus Sidonius shows in his verses what is but too apparent in the history of the elevation of Anthemius, that Nova Roma on the borders of Europe and Asia was the real sovereign.[2] And we also learn that the whole internal order of government, the structure of Roman law, and the daily habit of life had remained unaltered by barbarian occupation. This is the last time that Rome appears in garments of joy. The last reflection of her hundred triumphs still shines upon her palaces, baths, and temples. The Roman people, diminished in number, but unaltered in character, still frequented the baths of Nero, of Agrippa, of Diocletian; and Sidonius recommends instead baths less splendid, but less seductive to the senses.[3]

But Anthemius lasted no longer than the noble Majorian or the ignoble Severus. East and West had

---

[1] "Sed si forte placet veteres sopire querelas
Anthemium concede mihi; sit partibus istis
Augustus longumque Leo; mea jura gubernet
Quem petii."—*Carmen*, ii.

[2] Reumont, i. 700.

[3] He says at the end of 500 hendecasyllabics (jam te veniam loquacitati Quingenti hindecasyllabi precantur):

"Hinc ad balnea non Neroniana,
Nec quæ Agrippa dedit, vel ille cujus
Bustum Dalmaticæ vident Salonæ,
Ad thermas tamen ire sed libebat,
Privato bene præbitas pudori".

united their strength in a great expedition to put down the incessant Vandal piracies, which made all the coasts of the Mediterranean insecure.[1] It failed through the treachery of the eastern commander Basiliscus, to whose evil deeds we shall have hereafter to recur. This disaster shook the credit of Anthemius, and Ricimer also tired of his father-in-law. He went to Milan, and Rome was terrified with the report that he had made a compact with barbarians beyond the Alps. Ricimer marched upon Rome, to which he laid siege in 472. Here he was joined by Anicius Olybrius, who had married Placidia, the younger daughter of Valentinian and Eudoxia, through whom he claimed the throne, as representative of the Theodosian line. Ricimer, after a fierce contest with Anthemius, burst into the Aurelian gate at the head of troops all of German blood and Arian belief, massacring and plundering all but two of the fourteen regions. But the city escaped burning.

Then Anicius Olybrius entered Rome, consumed at once by famine, pestilence, and the sword. With the consent of Leo, and at the request of Genseric, he had been already named emperor. He took possession of the imperial palace, and made the senate acknowledge him. Anthemius had been cut in pieces, but forty days after his death Ricimer died of the plague, and thus had not been able to put to death more than four Roman emperors, of whom his father-in-law, Anthemius, was the last. The Arian Condottiere, who had inflicted

---

[1] For a well-told account of this expedition and its failure, see Thierry, *Derniers Temps de l'Empire d'Occident*, pp. 77-101.

on Rome a third plundering, said to be worse than that of Genseric, was buried in the Church of St. Agatha in Suburra,[1] which had been ceded to the Arians, and which he had adorned.

Olybrius made the Burgundian prince Gundebald commander of the forces, but died himself in October of that same year, 472, and left the throne to be the gift of barbarian adventurers. Three more shadows of emperors passed. Gundebald gave that dignity at Ravenna, in March, 473, to Glycerius, a man of unknown antecedents. In 474, Glycerius was deposed by Nepos, a Dalmatian, whom the empress Verina, widow of Leo I., had sent with an army from Byzantium to Ravenna. Nepos compelled his predecessor to abdicate, and to become bishop of Salona. He himself was proclaimed emperor at Rome on the 24th June, 474, after which he returned to Ravenna. While he was here treating with Euric, the Visigoth king, at Toulouse, Orestes, whom he had made Patricius and commander of the barbaric troops for Gaul, rose against him. Nepos fled by sea from Ravenna in August, 475, and betook himself to Salona, whither he had banished Glycerius.

Orestes was a Pannonian; had been Attila's secretary; then commander of German troops in service of the emperors. Thus he came to lead the troops which had been under Ricimer. This heap of Germans and

---

[1] There is a strange occurrence recorded by St. Gregory in his *Dialogues* as having taken place in this church, which would seem to point at Ricimer's burial in it.

Sarmatians without a country were in wild excitement, demanding a cession of Italian lands, instead of a march into Gaul. They offered their general the crown of Italy. Orestes thought it better to invest therewith his young son, and so, on the 31st October, 475, the boy Romulus Augustus, by the supremest mockery of what is called fortune, sat for a moment on the seat of the first king and the first emperor of Rome.

Italy could no longer produce an army, and the foreign soldiery who had served under various leaders naturally desired the partition of its lands. Odoacer was now their leader, who, when a penniless youth, had visited St. Severinus in Noricum, and received from him the prophecy: "Go into Italy, clad now in poor skins: thou wilt speedily be able to clothe many richly". Odoacer, after an adventurous life of heroic courage, made the homeless warriors whom he now commanded understand that it was better to settle on the fair lands of Italy than wander about in the service of phantom emperors. They acclaimed him as their king, and after beheading Orestes and getting possession of Romulus Augustus, he compelled him to abdicate before the senate, and the senate to declare that the western empire was extinct. This happened in the third year of the emperor Zeno the Isaurian, the ninth of Pope Simplicius, A.D. 476. The senate sent deputies to Zeno at Byzantium to declare that Rome no longer required an independent emperor; that one emperor was sufficient for East and for West; that they had chosen for the protector of Italy Odoacer, a man skilled in the arts of peace as well as war, and besought Zeno to entrust him with the dignity of Patricius and the government of Italy. The deposed Nepos also sent a petition to Zeno to restore him. Zeno replied to the senate that of the two emperors whom he had sent to them, they had deposed Nepos and killed Anthemius. But he received the diadem and the imperial jewels of the western empire, and kept them in his palace. He endured the usurper who had taken possession of Italy until he was able to put him down, and so, in his letters to Odoacer, invested him with the title of " Patricius of the Romans," leaving the government of Italy to a German commander under his imperial authority. So the division into East and West was cancelled : Italy as a province belonged still to the one emperor, who was seated at Byzantium. In theory, the unity of Constantine's time was restored ; in fact, Rome and the West were surrendered to Teuton invaders.[1] This was the last stroke : the mighty members of the great mother—Gaul, and Spain, and Britain, and Africa, and Illyricum—had been severed from her. Now, the head, discrowned and impotent, submitted to the rule of Odoacer the Herule. The Byzantine supremacy remained in keeping for future use. It had been acknowledged from the death of Honorius in 423, when Galla Placidia had become empress and her son emperor by the gift and the army of Theodosius II.

The agony of imperial Rome lasted twenty-one years. Valentinian III. was reigning in 455 : in the March of

---

[1] This account has been shortened from that of Gregorovius, i. 231-5.

that year he was murdered, and succeeded by Maximus, who was murdered in June; then by Avitus in July, who was murdered in October, 456. Majorianus followed in 457, and reigned till August, 461: he was followed by Libius Severus in November, who lasted four years, till November, 465. After an interregnum of eighteen months, in which Ricimer practically ruled, Anthemius was brought from Byzantium in April, 467, and continued till July, 472; but Anicius Olybrius again was brought from Byzantium, reigned for a few months in 472, and died of the plague in October. In 473, Glycerius was put up for emperor; in 474, he gave place to Nepos, the third brought from Byzantium. In 475, Romulus Augustus appears, to disappear in 476, and end his life in retirement at the Villa of Lucullus by Naples, once the seat of Rome's most luxurious senator.

Eighty years had now passed since the death of Theodosius. In the course of these years the realm which he had saved from dissolution after the defeat and death of Valens near Adrianople, and had preserved during fifteen years by wisdom in council and valour in war, and still more by his piety, when once his protecting hand and ruling mind were withdrawn, fell to pieces in the West, and was scarcely saved in the East. Let us take the last five years of St. Leo, which follow on the raid of Genseric, in order to complete the sketch just given of Rome's political state, by showing the condition of the great provinces which belonged to Leo's special patriarchate. I have before noticed how it was in the interval between the retirement of Attila from Rome at the prayer of St. Leo and the seizure of Rome by Genseric at the solicitation of the miserable empress Eudoxia, when St. Leo could save only the lives of his people, that he confirmed the Fourth Ecumenical Council. Not only was he entreated to do this by the emperor Marcian: the Council itself solicited the confirmation of its acts, which for that purpose were laid before him, while it made the most specific confession of his authority as the one person on earth entrusted by the Lord with His vineyard. From the particular time and the circumstances under which these events took place, one may infer a special intention of the Divine Providence. This was that the whole Roman empire, while it still subsisted, the two emperors, one of whom was on the point of disappearing, and the whole episcopate, in the most solemn form, should attest the Roman bishop's universal pastorship. For a great period was ending, the period of the Græco-Roman civilisation, from which, after three centuries of persecution, the Church had obtained recognition. And a great period was beginning, when the wandering of the nations had prepared for the Church another task. The first had been to obtain the conversion of nations linked by the bond of one temporal rule, enjoying the highest degree of culture and knowledge then existing, but deeply tainted by the corruption of effete refinement. The second was to exalt rough, sturdy, barbarian natures, whose bride was the sword and human life their prey, first to the virtues of the civil state, and next to the higher life of Christian charity, and thus to link them, who had known only violent repulsion and perpetual warfare among themselves, in not a temporal but a spiritual bond. The majestic figure of St. Leo expressed the completion of the first task. It also symbolises the beneficent power which in the course of ages will accomplish the second.

The wandering of the nations, says a great historian, was of decisive effect for the Church, and he quotes another historian's summary description of it: "It was not the migration of individual nomad hordes, or masses of adventurous warriors in continuous motion, which produced changes so mighty. But great, long-settled peoples, with wives and children, with goods and chattels, deserted their old seats, and sought for themselves in the far distance a new home. By this the position of individuals, of communities, of whole peoples, was of necessity completely altered. The old conditions of possession were dissolved. The existing bonds of society loosened. The old frontiers of states and lands passed away. As a whole city is turned into a ruinous heap by an earthquake, so the whole political system of previous times was overthrown by this massive transmigration. A new order of things had to be formed corresponding to the wholly altered circumstances of the nation."[1]

I draw from the same historian[2] an outline of the

---

[1] Giesebrecht, quoted by Hergenröther, *K.G.*, i. 449.
[2] Hergenröther, i. 449-453.

movement, running through several centuries, which had this final result. Great troops of Celts had, before the time of Christ, sought to settle themselves in Rhœtia and Upper Italy, even as far as Rome. Cimbrians and Teutons, with as little success, had betaken themselves southwards, while under the empire the pressure of peoples had more and more increased, and Trajan could hardly maintain the northern frontier on the Danube. In the third century, Alemans and Sueves advanced to the Upper Rhine, and the Goths, from dwelling between the Don and Theiss, came to the Danube and the Black Sea. Decius fell in battle with them. Aurelian gave them up the province of Dacia. Constantine the Great conquered them, and had Gothic troops in his army. Often they broke into the Roman territory, and carried off prisoners with them. Some of these were Christians and introduced the Goths to the knowledge of Christianity. Theophilus, a Gothic bishop, was at the Nicene Council in 325. They had clergy, monks, and nuns, with numerous believers. Under Athanarich, king of the Visigoths, Christians already suffered, with credit, a bloody persecution. On the occasion of the Huns, a Scythian people, compelling the Alans on the Don to join them, then conquering the Ostrogoths and oppressing the Visigoths, the latter prevailed on the emperor Valens to admit them into the empire. Valens gave them dwellings in Thrace on the condition that they should serve in his army and accept Arian Christianity. So the larger number of Visigoths under Fridiger in 375 became Arians. They soon, however, broke into conflict with the empire through their ill-treatment by the imperial commanders. In 378, Valens was defeated near Adrianople; his army was utterly crushed; he met himself with a miserable death. After this the Visigoths in general continued to be Arians, though many, especially through the exertions of St. Chrysostom, were converted to Catholicism. Most of them, however, seem to have been only half Arians, like their famous bishop Ulphilas. He was by birth a Goth—some say a Cappadocian—was consecrated between 341 and 348, in Constantinople. He gave the Goths an alphabet of their own, formed after the Greek, and made for them a translation of the Bible, of great value as a record of ancient German. He died in Constantinople before 388—probably in 381.

Under Theodosius I., about 382, the Visigoths accepted the Roman supremacy, and the engagement to supply 40,000 men for the service of the empire, upon the terms of occupying, as allies free of tribute, the provinces assigned to them of Dacia, Lower Mœsia, and Thrace. After this, discontented at the holding back their pay, and irritated by Rufinus, who was then at the head of the government of the emperor Arcadius, they laid waste the Illyrian provinces down to the Peloponnesus, and made repeated irruptions into Italy, in 400 and 402, under their valiant leader Alarich. In 408 he besieged Rome, and exacted considerable sums from it. He renewed the siege in 409, and made the wretched prefect Attalus emperor, whom he afterwards deposed, and recognised Honorius again. At last he took Rome by storm on the 24th August, 410. The city was completely plundered, but the lives of the people spared. He withdrew to Lower Italy and soon died. His brother-in-law and successor, Ataulf, was first minded entirely to destroy the Roman empire, but afterwards to restore it by Gothic aid. In the end he went to Gaul, conquered Narbonne, Toulouse, and Bordeaux, and afterwards Barcelona. His half-brother Wallia, after reducing the Alans and driving back the Sueves and Vandals, planted his seat in Toulouse, which became, in 415, the capital of his Aquitanean kingdom, Gothia or Septimania. Gaul, in which several Roman commanders assumed the imperial title, was overrun in the years from 406 to 416 by various peoples, whom the two opposing sides called in: by Burgundians, Franks, Alemans, Vandals, Quades, Alans, Gepids, Herules. The Alans, Sueves, Vandals, and Visigoths, at the same time, went to Spain. Their leaders endeavoured to set up kingdoms of their own all over Gaul and Spain.

Arianism came from the Visigoths not only to the Ostrogoths but also to the Gepids, Sueves, Alans, Burgundians, and Vandals. But these peoples, with the exception of the Vandals and of some Visigoth kings, treated the Catholic religion, which was that of their Roman subjects, with consideration and esteem. Only here and there Catholics were compelled to embrace Arianism. Their chief enemy in Gaul was the Visigoth king Eurich. Wallia, dying in 419, had been succeeded by Theodorich I. and Theodorich II., both of whom had extended the kingdom, which Eurich still more increased. He died in 483. Under him many Catholic churches were laid waste, and the Catholics suffered a bloody persecution. He was rather the head of a sect than the ruler of subjects. This, however, led to the dissolution of his kingdom, which, from 507, was more and more merged in that of the Franks.

The Burgundians, who had pressed onwards from the Oder and the Vistula to the Rhine, were in 417 already Christian. They afterwards founded a kingdom, with Lyons for capital, between the Rhone and the Saone. Their king Gundobald was Arian. But Arianism was not universal; and Patiens, bishop of Lyons, who died in 491, maintained the Catholic doctrine. A conference between Catholics and Arians in 499 converted few. But Avitus, bishop of Vienne, gained influence with Gundobald, so that he inclined to the Catholic Church, which his son Sigismund, in 517, openly professed. The Burgundian kingdom was united with the Frankish from 534.

The Sueves had founded a kingdom in Spain under their king Rechila, still a heathen. He died in 448. His successor, Rechiar, was Catholic. When king Rimismund married the daughter of the Visigoth king Theodorich, an Arian, he tried to introduce Arianism, and persecuted the Catholics, who had many martyrs—Pancratian of Braga, Patanius, and others. It was only between 550 and 560 that the Gallician kingdom of the Sueves, under king Charrarich, became Catholic, when his son Ariamir or Theodemir was healed by the intercession of St. Martin of Tours, and converted by Martin, bishop of Duma. In 563 a synod was held by the metropolitan of Braga, which established the Catholic faith. But in 585, Leovigild, the Arian king of the larger Visigoth kingdom, incorporated with his territory the smaller kingdom of the Sueves. Catholicism was still more threatened when Leovigild executed his own son Hermenegild, who had married the Frankish princess Jugundis, for becoming a Catholic. But the martyr's brother, Reccared, was converted by St. Leander, archbishop of Seville, and in 589 publicly professed himself a Catholic. This faith now prevailed through all Spain.

The Vandals, rudest of all the German peoples, had been invited by Count Boniface, in 429, to pass over from Spain under their king Genseric to the Roman province of North Africa. They quickly conquered it entirely. Genseric, a fanatical Arian, persecuted the Catholics in every way, took from them their churches, banished their bishops, tortured and put to death many. Some bishops he made slaves. He exposed Quodvultdeus, bishop of Carthage, with a number of clergy, to the mercy of the waves on a wretched raft. Yet they reached Naples. The Arian clergy encouraged the king in all his cruelties. It was only in private houses or in suburbs that the Catholics could celebrate their worship. The violence of his tyranny, which led many to doubt even the providence of God, brought the Catholic Church in North Africa into the deepest distress. Genseric's son and successor, Hunnerich, who reigned from 477 to 484, was at first milder. He had married Eudoxia, elder daughter of Valentinian III. The emperor Zeno had specially recommended to him the African Catholics. He allowed them to meet again, and, after the see of Carthage had been vacant twenty-four years, to have a new bishop. So the brave confessor Eugenius was chosen in 479. But this favour was followed by a much severer persecution. Eugenius, accused by the bitter Arian bishop Cyrila, was severely ill-treated, shut up with 4976 of the faithful, banished into the barest desert, wherein many died of exhaustion. Hunnerich stripped the Catholics of their goods, and banished them chiefly to Sardinia and Corsica. Consecrated virgins were tortured to extort from them admission that their own clergy had committed sin with them. A conference held at Carthage in 484 between Catholic and Arian bishops was made a pretext for fresh acts of violence, which the emperor Zeno, moved by Pope Felix III. to intercede, was unable to prevent. 348 bishops were banished. Many died of ill usage. Arian baptism was forced upon not a few, and very many lost limbs. This persecution produced countless martyrs. The greatest wonders of divine grace were shown in it. Christians at Tipasa, whose tongues had been cut out at the root, kept the free use of their speech, and sang songs of praise to Christ, whose godhead was mocked by the Arians. Many of these came to Constantinople, where the imperial court was witness of the miracle. The successor of this tyrant Hunnerich, king Guntamund, who reigned from 485 to 496, treated the Catholics more fairly, and, though the persecution did not entirely cease, allowed, in 494, the banished bishops to return. A Roman Council, in 487 or 488, made the requisite regulations with regard to those who had suffered iteration of baptism, and those who had lapsed. King Trasamund, from 496 to 523, wished again to make Arianism dominant, and tried to gain individual Catholics by distinctions. When that did not succeed, he went on to oppression and banishment, took away the churches, and forbade the consecration of new bishops. As still they did not diminish, he banished 120 to Sardinia, among them a great defender of the Catholic faith, St. Fulgentius, bishop of Ruspe. King Hilderich, who reigned from 523 to 530, a gentle prince and friend of the emperor Justinian, stopped the persecution and recalled the banished. Fulgentius was received back with great joy, and in February, 525, Archbishop Bonifacius held at Carthage a Council once more, at which sixty bishops were present. Africa had still able theologians. Hilderich was murdered by his cousin Gelimer: a new persecution was preparing. But the Vandal kingdom in Africa was overthrown in 533 by the eastern general Belisarius, and northern Africa united with Justinian's empire. However, the African Church never flourished again with its former lustre.

But Gaul and Italy had been in the greatest danger of suffering a desolation in comparison with which even the Vandal persecution in Africa would have been light. St. Leo was nearly all his life contemporaneous with the terrible irruptions of the Huns. These warriors, depicted as the ugliest and most hateful of the human race, in the years from 434 to 441, having already advanced, under Attila, from the depths of Asia to the Wolga, the Don, and the Danube, pressing the Teuton tribes before them, made incursions as far as Scandinavia. In the last years of the emperor Theodosius II. they filled with horrible misery the whole range of country from the Black Sea to the Adriatic. In the spring of 451 Attila broke out from Pannonia with 700,000 men, absorbed the Alemans and other peoples in his host, wasted and plundered populous cities such as Treves, Mainz, Worms, Spires, Strasburg, and Metz. The skill of Aetius succeeded in opposing him on the plains by Chalons with the Roman army, the Visigoths, and their allies. The issue of this battle of the nations was that Attila, after suffering and inflicting fearful slaughter, retired to Pannonia. The next year he came down upon Italy, destroyed Aquileia, and the fright of his coming caused Venice to be founded on uninhabited islands, which the Scythian had no vessels to reach. He advanced over Vicenza, Padua, Verona, Milan. Rome was before him, where the successor of St. Peter stopped him. He withdrew from Italy, made one more expedition against the Visigoths in Gaul, but died shortly after. With his death his kingdom collapsed. His sons fought over its division, the Huns disappeared, and what was afterwards to be Europe became possible.

The invasions of the Hun shook to its centre the western empire. Aetius, who had saved it at Chalons in 451, received in 454 his death-blow as a reward from the hand of Valentinian III., and so we are brought to the nine phantom emperors who follow the race of the great Theodosius, when it had been terminated by the vice of its worst descendant.

One Teuton race, the most celebrated of all, I have reserved for future mention. The Franks in St. Leo's time, and for thirty-five years after his death, were still pagan. The Salian branch occupied the north of Gaul, and the Ripuarians were spread along the Rhine, about Cologne. Their paganism had prevented them from being touched by the infection of the Arian heresy, common to all the other tribes, so that the Arian religion was the mark of the Teutonic settler throughout the West, and the Catholic that of the Roman provincials.

Thus when, in the year 476, the Roman senate, at Odoacer's bidding, exercised for the last time its still legal prerogative of naming the emperor, by declaring that no emperor of the West was needed, and by sending back the insignia of empire to the eastern emperor Zeno, all the provinces of the West had fallen, as to government, into the hands of the Teuton invaders, and all of these, with the single exception of the Franks, were Arians. They alone were still pagans. Odoacer, also an Arian, became the ruler of Rome and Italy, nominally by commission from the emperor Zeno, really in virtue of the armed force, consisting of adventurers belonging to various northern tribes which he commanded. To the Romans he was Patricius,[1] a title of honour lasting for life, which from Constantine's time,

---

[1] Reumont, ii. 6.

without being connected with any particular office, surpassed all other dignities. To his own people he was king of the Ruges, Herules, and Turcilings, or king of the nations. He ruled Italy, and Sicily, except a small strip of coast, and Dalmatia, and these lands he was able to protect from outward attack and inward disturbance. He made Ravenna his seat of government. He did not assume the title of king at Rome. He maintained the old order of the State in appearance. The senate held its usual sittings. The Roman aristocracy occupied high posts. The consuls from the year 482 were again annually named. The Arian ruler left theological matters alone. But the eyes of Rome were turned towards Byzantium. The Roman empire continued legally to exist, and especially in the eye of the Church. The Pope maintained relations with the imperial power.

In the meantime, Theodorich the Ostrogoth, son of Theodemir, chief of the Amal family, had been sent as a hostage for the maintenance of the treaty made by the emperor Leo I. with his father, and had spent ten years, from his seventh to his seventeenth year, at Constantinople. Though he scorned to receive an education in Greek or Roman literature, he studied during these years, with unusual acuteness, the political and military circumstances of the empire. Of strong but slender figure, his beautiful features, blue eyes with dark brows, and abundant locks of long, fair hair, added to the nobility of his race, pointed him out for a future ruler.[1]

[1] Reumont, ii. 9.

In 475, Theodorich succeeded his father as king of the Ostrogoths in their provinces of Pannonia and Mœsia, which had been ceded by the empire. He it was who was destined to lead his people to glory and greatness, but also to their fall, in Italy. Zeno had striven to make him a personal friend—had made him general, given him pay and rank. Theodorich had not a little helped Zeno in his struggle for the empire. The Ostrogoth, in 484, became Roman consul; but he also appeared suddenly in a time of peace before the gates of Constantinople, in 487, to impress his demands upon Zeno. Theodorich and his people occupied towards Zeno the same position which Alaric and his Visigoths had held towards Honorius. Their provinces were exhausted, and they wanted expansion. Whether it was that Zeno deemed the Ostrogothic king might be an instrument to terminate the actual independence of Italy from his empire, or that the neighbourhood of the Goths, under so powerful a ruler, seemed to him dangerous, or that Theodorich himself had cast longing eyes upon Italy, Zeno gave a hesitating approval to the advance of the last great Gothic host to the southwest. The first had taken this direction under Alaric eighty-eight years before. Now a sovereign sanction from the senate of Constantinople, called a Pragmatic sanction, assigned Italy to the Gothic king and his people.

From Novæ, Theodorich's capital on the Danube, not far from the present Bulgarian Nikopolis, this world of wanderers, numbered by a contemporary as at least

350,000, streamed forth with its endless train of waggons. At the Isonzo, Italy's frontier, Odoacer, on the 28th August, 489, encountered the flood, and was worsted, as again at the Adige. Then he took refuge in Ravenna. The end of a three years' conflict, in which the Gothic host was encamped in the pine-forest of Ravenna, and where the "Battle of the Ravens" is commemorated in the old German hero-saga, was that, in the winter of 493, the last refuge of Odoacer opened its gates. Odoacer was promised his life, but the compact was broken soon. His people proclaimed Theodorich their king. Theodorich had sent a Roman senator to Zeno to ask his confirmation of what he had done. Zeno had been succeeded by Anastasius in 491. How much Anastasius granted cannot be told. Rome, during this conflict, had remained in a sort of neutrality. At first Theodorich deprived of their freedom as Roman citizens all Italians who had stood in arms against him. Afterwards, he set himself to that work of equal government for Italians and Goths which has given a lustre to his reign, though the fair hopes which it raised foundered at last in an opposition which admitted of no reconcilement.

Theodorich[1] reigned from 493 to 526. He extended by successful wars the frontiers of the Gothic kingdom beyond the mainland of Italy and its islands. Narbonensian Gaul, Southern Austria, Bosnia, and Servia belonged to it at its greatest extension. The Theiss and

---

[1] Reumont (ii. 29-42) gives an admirable sketch of the government of Theodorich, by which I have profited in what follows.

the Danube, the Garonne and the Rhone, flowed beside his realm. The forms of the new government, as well as the laws, remained the same substantially as in Constantine's time. The Roman realm continued, only there stood at its head a foreign military chief, surrounded by his own people in the form of an army. Romandom lived on in manner of life, in customs, in dress. The Romans were judged according to their own laws. Gothic judges determined matters which concerned the Goths; in cases common to both they sat intermixed with Roman judges. Theodorich's principle was with firm and impartial hand to deal evenly between the two. But the military service was reserved to the Goths alone. Natives were forbidden even to carry knives. The Goths were to maintain public security: the Romans to multiply in the arts of peace. But even Theodorich could not fuse these nations together. The Goths remained foreigners in Italy, and possessed as *hospites* the lands assigned to them, which would seem to have been a third. This noblest of barbarian princes, and most generous of Arians, had to play two parts. In Ravenna and Verona he headed the advance of his own people, and was king of the Goths: in Rome the Patricius sought to protect and maintain. When, in 500, he visited Rome, he was received before its gates by the senate, the clergy, the people, and welcomed like an emperor of the olden time. Arian as he was, he prayed in St. Peter's, like the orthodox emperors of the line of Theodosius, at the Apostle's tomb. Before the senate-house, in the forum, Boethius greeted him with a speech. The German king admired the forum of Trajan, as the son of Constantine, 143 years before, had admired it. Statues in the interval had not ceased to adorn it. Romans and Franks, heathens and Christians, alike were there: Merobaudes, the Gallic general; Claudian, the poet from Egypt, the worshipper of Stilicho, in verses almost worthy of Virgil; Sidonius Apollinaris, the future bishop of Clermont, who panegyrised three emperors successively deposed and murdered. The theatre of Pompey and the amphitheatre of Titus still rose in their beauty; and as the Gothic king inhabited the vast and deserted halls of the Cæsarean palace, he looked down upon the games of the Circus Maximus, where the diminished but unchanged populace of Rome still justified St. Leo's complaint, that the heathen games drew more people than the shrines of the martyrs whose intercession had saved Rome from Attila. In fine, St. Fulgentius could still say, If earthly Rome was so stately, what must the heavenly Jerusalem be!

The bearing of the Arian king to the Catholic Church and the Roman Pontificate was just and fair almost to the end of his reign. He protected Pope Symmachus at a difficult juncture. His minister Cassiodorus supported and helped the election of Pope Hormisdas. The letters of Cassiodorus, as his private secretary, counsellor, and intimate friend, remain to attest, with the force of an eye-witness, a noble Roman and a devoted Christian, who was also Patricius and Prætorian Prefect—the nature of the government, as well as the state of Italian society at that time. We hardly possess such another source of knowledge for this century. But under Pope John I. this happy state of things broke down. A dark shadow has been thrown upon the last years of an otherwise glorious government. The noble Boethius, after being leader of the Roman senate and highly-prized minister of the Gothic king, died under hideous torture, inflicted at the command of a suspicious and irritated master. Again, he had forced upon Pope John I. an embassy to Constantinople, and required of him to obtain from the eastern emperor churches for Arians in his dominions. The Pope returned, after being honoured at the eastern court as the first bishop of the world, laden with gifts for the churches at Rome, but without the required consent of the emperor to give churches to the Arians. He perished in prison at Ravenna by the same despotic command. This was in May, 526, and in August the king himself died almost suddenly, fancying, it was said, that he saw on a fish which was brought to his table the head of a third victim, the illustrious Symmachus. What Catholics thought of his end is shown by St. Gregory seventy years afterwards, who records in his Dialogues a vision seen at Lipari on the day of the king's death, in which the Pope and Symmachus were carrying him between them with his hands tied, to plunge him in the crater of the volcano.

Several writers[1] have termed Theodorich a premature

[1] Montalembert, Gregorovius, Kurth. Philips (vol. iii., p. 51, sec. 119), remarks: "Wäre Theodorich der Grosse nicht Arianer gewesen, so würde, wenn er es sonst gewollt, ihm wohl nichts weiter im Wege gestanden haben, als sich zum Römischen kaiser im Abendlande ansrufen lassen".

Charlemagne. It seems to me that, as Genseric was the worst and most ignoble of the Teutonic Arian princes, Theodorich was the best. The one showed how cruel and remorseless an Arian persecutor was, the other how fair a ruler and generous a protector the nature of things would allow an Arian monarch to be. But in his case the end showed that the Gothic dominion in Italy rested only on the personal ability of the king, and, further, that no stable union could take place until these German-Arian races had been incorporated by the Catholic Church into her own body.[1]

This truth is yet more illustrated by a double contrast between Theodorich and Clovis. In personal character the former was far superior to the latter. Clovis was converted at the age of thirty, and died at forty-five. Yet the effect of the fifteen years of his reign after he became a Catholic was permanent. From that moment the Franks became a power. In that short time Clovis obtained possession of a very great part of France, and that possession went on and was confirmed to his line and people. The thirty-three years of Theodorich secured to Italy a time of peace, even of glory, which did not fall to its lot for ages afterwards. Yet the effect of his government passed with him; his daughter and heiress, the noble princess Amalasuntha, in whose praise Cassiodorus exhausts himself, was murdered; his kingdom was broken up, and Cassiodorus himself, retiring from public life, confessed in his monastic life, continued for a generation, how vain had been the attempt of the Arian

[1] Gregorovius, i. 312, 315.

king to overcome the antagonistic forces of race and religion by justice, valour, and forbearance.

It was fitting that the attempt should be made by the noblest of Teutonic races, under the noblest chief it ever produced. Nor is it unfitting here to recur to the opinion of another great Goth, not indeed the equal of Theodorich, yet of the same race and the nearest approach to him, one of those conquerors who showed a high consideration for the Roman empire. Orosius records " that he heard a Gallic officer, high in rank under the great Theodosius, tell St. Jerome at Bethlehem how he had been in the confidence of Ataulph, who succeeded Alaric, and married Galla Placidia. How he had heard Ataulph declare that, in the vigour and inexperience of youth, he had ardently desired to obliterate the Roman name, and put the Gothic in its stead—that instead of Romania the empire should be Gothia, and Ataulph be what Augustus had been. But a long experience had taught him two things—the one, that the Goths were too barbarous to obey laws; the other, that those laws could not be abolished, without which the commonwealth would cease to be a commonwealth. And so he came to content himself with the glory of restoring the Roman name by Gothic power, that posterity might regard him as the saviour of what he could not change for the better."[1]

It seems that the observation of Ataulph at the beginning of the fifth century was justified by the experience of Theodorich at the beginning of the sixth.

[1] Orosius, *Hist.*, vii. 43.

And, further, we may take the conduct of these two great men as expressing on the whole the result of the Teutonic migration in the western provinces. After unspeakable misery produced in the cities and countries of the West at the time of their first descent, we may note three things. The imperial lands, rights, and prerogatives fell to the invading rulers. The lands in general partly remained to the provincials (the former proprietors), partly were distributed to the conquerors. But for the rest, the fabric of Roman law, customs, and institutions remained standing, at least for the natives, while the invaders were ruled severally according to their inherited customs. Even Genseric was only a pirate, not a Mongol, and after a hundred years the Vandal reign was overthrown and North Africa reunited to the empire. In the other cases it may be said that the children of the North, when they succeeded, after the struggle of three hundred years, in making good their descent on the South, seized indeed the conqueror's portion of houses and land, but they were not so savage as to disregard, in Ataulph's words, those laws of the commonwealth, without which a commonwealth cannot exist. The Franks, in their original condition one of the most savage northern tribes, in the end most completely accepted Roman law, the offspring of a wisdom and equity far beyond their power to equal or to imitate. And because they saw this, and acted on it most thoroughly, they became a great nation. The Catholic faith made them. Thus, when the boy Romulus Augustus was deposed at

Rome, and power fell into the hands of the Herule Odoacer, Pope Simplicius, directing his gaze over Africa, Spain, France, Illyricum, and Britain, would see a number of new-born governments, ruled by northern invaders, who from the beginning of the century had been in constant collision with each other, perpetually changing their frontiers. Wherever the invaders settled a fresh partition of the land had to be made, by which the old proprietors would be in part reduced to poverty, and all the native population which in any way depended on them would suffer greatly. It may be doubted whether any civilised countries have passed through greater calamities than fell upon Gaul, Spain, Eastern and Western Illyricum, Africa, and Britain in the first half of the fifth century. Moreover, while one of these governments was pagan, all the rest, save Eastern Illyricum, were Arian. That of the Vandals, which had occupied, since 429, Rome's most flourishing province, also her granary, had been consistently and bitterly hostile to its Catholic inhabitants. That of Toulouse, under Euric, was then persecuting them. Britain had been severed from the empire, and seemed no less lost to the Church, under the occupation of Saxon invaders at least as savage as the Frank or the Vandal. In these broad lands, which Rome had humanised during four hundred years, and of which the Church had been in full possession, Pope Simplicius could now find only the old provincial nobility and the common people still Catholic. The bishops in these several provinces were exposed everywhere to an Arian succession of antagonists, who used against them all the influence of an Arian government.

When he looked to the eastern emperor, now become in the eyes of the Church the legitimate sovereign of Rome, by whose commission Odoacer professed to rule, instead of a Marcian, the not unworthy husband of St. Pulcheria, instead of Leo I., who was at least orthodox, and had been succeeded by his grandson the young child Leo II., he found upon the now sole imperial throne that child's father Zeno. He was husband of the princess Ariadne, daughter of Leo I.,[1] a man of whom the Byzantine historians give us a most frightful picture. Without tact and understanding, vicious, moreover, and tyrannical, he oppressed during the two years from 474 to 476 his people, sorely tried by the incursions of barbarous hordes. He also favoured, all but openly, the Monophysites, specially Peter Fullo, the heretical patriarch of Antioch. After two years a revolution deprived him of the throne, and exalted to it the equally vicious Basiliscus—the man whose treachery as an eastern general had ruined the success of the great expedition against Genseric, in which East and West had joined under Anthemius. Basiliscus still more openly favoured heresy. He lasted, however, but a short time; Zeno was able to return, and occupied the throne again during fourteen years, from 477 to 491. These two men, Zeno and Basiliscus, criminal in their private lives, in their public lives adventurers, who gained the throne by the worst Byzantine arts,

[1] Photius, i. 111.

opened the line of the theologising emperors. Basiliscus, during the short time he occupied the eastern throne, issued, at the prompting of a heretic whom he had pushed into the see of St. Athanasius—and it is the first example known in history—a formal decree upon faith, the so-called Encyclikon, in which only the Nicene, Constantinopolitan, and Ephesine Councils were accepted, but the fourth, that of Chalcedon, condemned. So low was the eastern Church already fallen that not the Eutycheans only, but five hundred Catholic bishops subscribed this Encyclikon, and a Council at Ephesus praised it as divine and apostolical.

Basiliscus, termed by Pope Gelasius the tyrant and heretic, was swept away. But his example was followed in 482 by Zeno, who issued his Henotikon, drawn up it was supposed by Acacius of Constantinople,[1] addressed to the clergy and people of Alexandria. Many of the eastern bishops, through fear of Zeno and his bishop Acacius, submitted to this imperial decree; many contended for the truth even to death against it. These two deeds, the Encyclikon of Basiliscus and the Henotikon of Zeno, are to be marked for ever as the first instances of the temporal sovereign infringing the independence of the Church in spiritual matters, which to that time even the emperors in Constantine's city had respected.

Simplicius sat in the Roman chair fifteen years, from 468 to 483; and such was the outlook presented to him in the East and West—an outlook of ruin, calamity,

---

[1] Photius, i. 120.

and suffering in those vast provinces which make our present Europe—an outlook of anxiety with a prospect of ever-increasing evil in the yet surviving eastern empire. There was not then a single ruler holding the Catholic faith. Basiliscus and Zeno were not only heretical themselves, but they were assuming in their own persons the right of the secular power to dictate to the Church her own belief. And the Pope had become their subject while he was locally subject to the dominion of a northern commander of mercenaries, himself a Herule and an Arian. In his own Rome the Pope lived and breathed on sufferance. Under Zeno he saw the East torn to pieces with dissension; prelates put into the sees of Alexandria and Antioch by the arm of power; that arm itself directed by the ambitious spirit of a Byzantine bishop, who not only named the holders of the second and third seats of the Church, but reduced them to do his bidding, and wait upon his upstart throne. Gaul was in the hand of princes, mostly Arian, one pagan. Spain was dominated by Sueves and Visigoths, both Arian. In Africa Simplicius during forty years had been witness of the piracies of Genseric, making the Mediterranean insecure, and the cities on every coast liable to be sacked and burnt by his flying freebooters, while the great church of Africa, from the death of St. Augustine, had been suffering a persecution so severe that no heathen emperor had reached the standard of Arian cruelty. In Britain, civilisation and faith had been alike trampled out by the northern pirates Hengist and

Horsa, and successive broods of their like. The Franks, still pagan, had advanced from the north of Gaul to its centre, destroyers as yet of the faith which they were afterwards to embrace. What did the Pope still possess in these populations? The common people, a portion of the local proprietors, and the Catholic bishops who had in him their common centre, as he in them men regarded with veneration by the still remaining Catholic population.

In all this there is one fact so remarkable as to claim special mention. How had it happened that the Catholic faith was considered throughout the West the mark of the Roman subject; and the Arian misbelief the mark of the Teuton invader and governor? Theodosius had put an end to the official Arianism of the East, which had so troubled the empire, and so attacked the Primacy in the period between Constantine and himself. During all that time the Arian heresy had no root in the West. But the emperor Valens, when chosen as a colleague by his brother Valentinian I., in 364, was counted a Catholic. A few years later he fell under the influence of Eudoxius, who had got by his favour the see of Byzantium. This man, one of the worst leaders of the Arians, taught and baptised Valens, and filled him with his own spirit; and Valens, when he settled the Goths in the northern provinces by the Danube, stipulated that they should receive the Arian doctrine. Their bishop and great instructor Ulphilas had been deceived, it is said, into believing that it was the doctrine of the Church. This fatal gift of a spurious doctrine the Goth received in all the energy of an uninstructed but vigorous will. As the leader of the northern races he communicated it to them. A Byzantine bishop had poisoned the wells of the Christian faith from which the great new race of the future was to drink, and when Byzantium succeeded in throwing Alaric upon the West, all the races which followed his lead brought with them the doctrine which Ulphilas had been deceived into propagating as the faith of Christ. So it happened that if the terrible overthrow of Valens in 378 by the nation which he had deceived brought his persecution with his reign to an end in the East, yet through his act Arianism came into possession, a century later, of all but one of the newly set up thrones in the West.

In truth, at the time the western empire fell the Catholic Church was threatened with the loss of everything which, down to the time of St. Leo, she had gained. For the triumph which Constantine's conversion had announced, for the unity of faith which her own Councils had maintained from Nicæa to Chalcedon, she seemed to have before her subjection to a terrible despotism in the East, extinction by one dominant heresy in the West. For here it was not a crowd of heresies which surrounded her, but the secular power at Rome, at Carthage, at Toulouse and Bordeaux, at Seville and Barcelona, spoke Arian. Who was to recover the Goth, the Vandal, the Burgundian, the Sueve, the Aleman, the Ruge, from that fatal error? Moreover, her bounds had receded. Saxon and Frank had largely swept away the Christian faith in their respective conquests. Who was to restore it to them? The Rome which had planted her colonies through these vast lands as so many fortresses, first of culture and afterwards of faith, was now reduced to a mere *municipium* herself. The very senate, with whose name empire had been connected for five hundred years, at the bidding of a barbarous leader of mercenaries serving for plunder, sent back the symbols of sovereignty to the adventurer, whoever he might be, who sat by corruption or intrigue on the seat of Constantine in Nova Roma.

This thought leads me to endeavour more accurately to point out the light thrown upon the Papal power by the various relations in which it stood at different times to the temporal governments with which it had to deal.

The practical division of the Roman empire in the fourth century, ensuing upon the act of Constantine in forming a new capital of that empire in the East, made the Church no longer subject to one temporal government. The same act tested the spiritual Primacy of the Church. It called it forth to a larger and more complicated action. I have in a former volume followed at considerable length the series of events the issue of which was, after Arian heretics had played upon eastern jealousy and tyrannical emperors during fifty years, to strengthen the action of the Primacy. But assuredly had that Primacy been artificial, or made by man, the division of interests ensuing upon the political disjunction of the East and West would have destroyed it. Julius and Liberius and Damasus would not have stood against Constantius and Valens if the heart of the Church had not throbbed in the Roman Primacy. Still more apparent does this become in the next fifty years, wherein the overthrow of the western empire begins. Then the sons of Theodosius, instead of joining hand with hand and heart with heart against the forces of barbarism, which their father had controlled and wielded, were seduced by their ministers into antagonism with each other. Byzantium worked woe to the elder sister of whom she was jealous. Under the infamous treasons of Rufinus and Eutropius, the words might have been uttered with even fuller truth than in their original application—

"Suis et ipsa Roma viribus ruit".

Thus Alaric first took Rome. But he did not take the Primacy. Pope Innocent lost no particle of his dignity or influence by the violation of Rome's secular dignity. It was only seven years after that event when St. Augustine and the two great African Councils acknowledged his Principate in the amplest terms. The heresy of Pelagius and the schism of Donatus were stronger than the sword of Alaric. And only a few years later, when a most fearful heresy, broached by the Byzantine bishop, led to the assembly in which then for the first time the Church met in general Council since Nicæa, the most emphatic acknowledgment of the Primacy as seated in the Roman bishop by descent from Peter was given by bishops, the subjects of an emperor very jealous of the West, to a Pope who could not live securely in Rome itself.

In all these hundred years it is seen how the division of the empire enlarged and strengthened the action of the Primacy. But this it did because the Primacy was divine. The events just referred to, but described elsewhere at length, would have destroyed it had it not been divine.

But this course of things, which is seen in action from the Nicene to the Chalcedonic Council, comes out with yet stronger force from the moment when Rome loses all temporal independence. We may place this moment at the date of its capture by Genseric. But it continues from that time. The events which took place at Rome in the twenty-one following years, the nine sovereigns put up and deposed, the subjection to barbarous leaders of hireling free-lances, the worse plundering of Ricimer seventeen years after that of Genseric —these were events grieving to the heart St. Leo and his successors; but yet not events at Rome alone— the whole condition of things in East and West which Pope Simplicius had to look upon outside of his own city, despotic emperors in the East, with bishops bending to their will, allowing the apostolic hierarchy to be displaced, and the apostolic doctrine determined by secular masters; Teuton settlements in the West ruled by the heresy most inimical to the Church; the Catholic population reduced in numbers and lowered in social position; whole countries seized by pagans, and forced at once into barbarism and infidelity—in the midst of all these the Pope stood: his generals were the several bishops of captured cities, whose places were assaulted by heretical rivals, supported by their kings. Gaul, Spain, Britain, Africa, Illyricum, Italy itself, no longer parts of one government, but ruled by enemies, any or all of these would have rejected the Roman Primacy if it had not come to them with the strongest warrant both of the Church's past history and her present consciousness.

Such was the new world in which the Pope stood from the year 455; and he stood in it for three hundred years. The testimony which such times bear is a proof superadded to the words of Fathers and the decrees of Councils.

But there is one other point in the political situation on which a word must be said.

From the time named, the Roman Primacy is the one sole fixed point in the West. All else is fluctuating and transitional. To the Pope the bishops, subject in each city to barbarian insolence, cling as their one unfailing support. Without him they would be Gothic, or Vandal, or Burgundian, or Sueve, or Aleman, or Turciling,— with him and in him they are Catholic. Let me express, in the words of another, what is contained in this fact. The Church, says Guizot, "at the commencement of the fifth century, had its government, a body of clergy, a hierarchy, which apportioned the different functions of the clergy, revenues, independent means of action, rallying points which suit a great society, councils provincial, national, general, the habit of arranging in common the society's affairs. In a word, at this epoch Christianity was not only a religion but a Church. If it had not been a Church, I do not know what would have become of it in the midst of the Roman empire's fall. I confine myself to purely human considerations: I put aside every element foreign to the natural consequences of natural facts. If Christianity had only been a belief, a feeling, an individual conviction, we may suppose that it would have broken down at the dissolution of the empire and the barbarian invasion. It did break down later in Asia and in all north Africa beneath an invasion of the same kind—that of barbarous Mussulmans. It broke down then though it was an institution, a constituted Church. Much more might the same fact have happened at the moment of the Roman empire's fall. There were then none of those means by which in the present day moral influences are established or support themselves independent of institutions: no means by which a naked truth, a naked idea, acquires a great power over minds, rules actions, and determines events. Nothing of the kind existed in the fourth century to invest ideas and personal feelings with such an authority. It is clear that a strongly organised, a strongly governed, society was needed to struggle against so great a disaster, to overcome such a hurricane. I think I do not go too far in affirming that, at the end of the fourth and the beginning of the fifth century, it is the Christian Church which saved Christianity. It is the Church, with its institutions, its magistrates, and its power, which offered a vigorous defence to the internal dissolution of the empire, to barbarism; which conquered the **barbarians**; which became the bond, the means, the **principle of** civilisation to the Roman and the barbarian world." [1]

In this passage, Guizot speaks of the Church as a government, as a unity. At the very moment of which he speaks, St. Augustine was addressing the Pope as the fountainhead of that unity; **and in the** midst of the dissolution an emperor **was recommending** him to the Gallic bishops " **as the chief of the** episcopal coronet" [2] encircling the earth. The whole structure which lasted through this earthquake of nations **had** its cohesion in him—a fact seen even more clearly in the time of the third Valentinian **than** in that of the conquering Constantine.

But looking to that **East**, which dates from the Encyclikon of a Basiliscus and the **Henotikon of a** Zeno, here the Pope appears as the sole check **to a** despotic power. He alone could speak to the **emperor on** an equal and even a superior footing. Would such a power not have repudiated his interference, **had it** not been convinced of an authority **beyond its reach to** deny? The first generation following the utter impotence of Rome as reduced to a *municipium* under Arian **rulers** will answer this question, as we shall see hereafter, with fullest effect.

I have adduced above three political situations. The first is when the Primacy passes from dealing with one government to deal with more than one; the second

---

[1] Guizot, *Sur la Civilisation en Europe*, deuxième leçon.
[2] Edict of Valentinian III., in 447.

when the Primacy has to deal with an unsettled world of many governments; the third when it is the sole fixed point in the face of a hurricane on one side and a despotism on the other. I observe that the testimony of all three concurs to bring out its action and establish its divine character. As an epilogue to all that has been said, I will suppose a case.

Three men, great with the natural greatness of intellect, greater still in the acquired greatness of character, greatest of all in the supernatural grace of saintliness, witnessed this fifth century from its beginning: one of them, during two decades of years; the second, during three; the last, during six decades. They saw in their own persons, or they heard in authentic narratives, all its doings—the cities plundered and overthrown; the countries wasted; all natural ties disregarded; neither age, nor sex, nor dignity, respected by hordes of savages, incapable themselves of learning, strangers to science, without perception of art; the sum being that the richest civilisation which the world had borne was crushed down by brute force. They saw, and mourned, and bore with unfailing personal courage their portion of sorrow, mayhap turning themselves in their inmost mind from a world perishing before their eyes, to contemplate the joy promised in a world which should not perish. But neither to St. Jerome, nor to St. Augustine, nor to St. Leo, did the thought occur that this barbarian mass could be controlled into producing a civilisation richer than that which its own incursion destroyed. That, instead of perpetual strife and mutual repulsion, it could receive the one law of Christ; be moulded into a senate of nations, with like institutions and identical principles; that, instead of one empire taking an external impress of the Christian faith, but rebelling against it with a deep-seated corruption and an unyielding paganism, and so perishing in the midst of abundance, it should grow into peoples, the corner-stone of whose government and the parent of their political constitution should be the one faith of Christ, and their acknowledged judge the Roman Pastor; and that the Rome which all the three saw once plundered, and the third twice subjected to that penalty, should lose all its power as a secular capital, while it became the shrine whence a divine law went forth; and that these hordes, who laid it waste before their eyes, should become its children and its most valiant defenders.

Had such a vision been vouchsafed to either of these great saints, with what words of thankfulness would he have described it. This is the subject which this narrative opens; and we, the long-descended offspring of these hordes, have seen this sight and witnessed this exertion of power carried on through centuries; and degenerate and ungrateful children as we are, we are living still upon the deeds which God wrought in that conversion of the nations by the pastoral staff of St. Peter, leading them into a land flowing with oil and wine.

CHAPTER II.

CÆSAR FELL DOWN.

WHEN St. Leo refused his assent to the Canons in favour of the see of Constantinople, which, at the end of the Council of Chalcedon, the Court, the clergy, and above all Anatolius, the bishop of the imperial city, desired to be passed, and with that intent overbore the resistance of the Papal legates, the race of Theodosius was still reigning both at Old and at New Rome. The eastern sovereigns, Marcian and Pulcheria, by becoming whose husband Marcian had ascended the throne, had acted with conspicuous loyalty towards the Pope. The mistakes of Theodosius II. were repaired, and the cabals of his courtiers ceased to affect the stronger minds and faithful hearts of his successors. In the West, Galla Placidia, during all the reign, since the death, in 423, of her brother Honorius, with which her nephew Theodosius II. had invested her, was also faithful to St. Peter's See; the same spirit directed her son Valentinian, and his empress-cousin, the daughter of the eastern emperor. The letters of all exist, in which they strove to set right their father, or nephew, Theodosius II., in the matter of Eutyches. All had supported St. Leo in the annulling that unhappy Council which compromised the faith of the Church so long as it was allowed to count as a Council. But not for any merit on the part of Pulcheria and Marcian would St. Leo allow the mere grandeur of a royal city, because it was the seat of empire, to dethrone from their original rank, held since the beginning of the Christian hierarchy, the two other Sees of St. Peter—the one of his disciple St. Mark, sent from his side at Rome; the other, in which he had first sat himself. St. Leo could not the least foresee that the course of things in less than a generation would justify by the plainest evidence of facts his maintenance of tradition and his prescience of future dangers. He had charged Anatolius with seeking unduly to exalt himself at the expense of his brethren. The exaltation consisted in making himself the second bishop of the Church. His see, a hundred and twenty years before, had, if it existed at all—for it is all but lost in insignificance—been merely a suffragan of the archbishop of Heraclea. Leo saw that Anatolius, under cover of the emperor's permanent residence in Nova Roma, sought to make its bishop the lever by which the whole episcopate of the East should be moved. We are now to witness the attempt to carry into effect all which St. Leo feared by a bishop who was next successor but one to Anatolius in his see.

The changes, indeed, wrought in a few years were immense. St. Leo himself outlived both Pulcheria and Marcian; and on the death of the latter saw the imperial succession, which had been in some sense hereditary since the election of Valentinian I., in 364, pass to a new man. As this is the first occasion on which the succession to the Byzantine throne comes into our review, it may be well to consider what sort of thing it was. I suppose the Cæsarean succession even from the first is a hard thing to bring under any definition. Since Claudius was discovered quaking for fear behind a curtain, and dragged out to sit upon the throne which his nephew Caius had hastily vacated, after having been welcomed to it four years before with universal acclamation, it would be difficult to say what made a man emperor of the Romans. So much I seem to see in that terrible line, that the descent from father to son was hardly ever blessed, and that those who were adopted by an emperor no way related to them succeeded the best. The children of the very greatest emperors—of a Marcus Aurelius, a Constantine, a Theodosius—have only brought shame on their parents and ruin on their empire. Again, if the youth of a Nero or a Caracalla ended in utter ignominy, the youth of an Alexander Severus produced the fairest of reigns, while it ended in his murder by an usurper. But strange and anomalous as the Cæsarean succession appears, that of the Byzantine sovereigns, from the disappearance of the Theodosian race to the last Constantine who dies on the ramparts of the city made by the first, shows a great deterioration.[1] There was no acknowledged principle of succession. Arbitrary force determined it. One robber followed another upon the throne; so that

[1] See Phillips, *Kirchenrecht*, vol. iii., sec. 119.

the eastern despot seemed to imitate that ghastly rule, in the wood by Nemi, " of the priest who slew the slayer and shall himself be slain ". If the army named one man to the throne, the fleet named another. If intrigue and shameless deceit gained it in one case, murder succeeded in another. Relationship or connection by marriage with the last possessor helped but rarely. This frequent and irregular change, and the personal badness of most sovereigns, caused endless confusion to the realm. This is the staple of the thousand years in which the election of the emperor Leo I., in 457, stands at the head. On the death of Marcian, following that of Pulcheria, in whose person a woman first became empress regnant, Leo was a Thracian officer, a colonel of the service, and director of the general Aspar's household. Aspar was an Arian Goth, commander of the troops, who had influence enough to make another man emperor, but not to cancel the double blot of barbarian and heretic in his own person. He made Leo, with the intention to be his master. And Leo ruled for seventeen years with some credit; and presently put Aspar and his son to death, in a treacherous manner, but not without reason. He bore a good personal character, was Catholic in his faith, and St. Leo lived on good terms with him during the four years following his election. St. Leo, dying in 461, was succeeded by Pope Hilarus, the deacon and legate who brought back a faithful report to Rome of the violent Council at Ephesus, in 449, from which he had escaped. Pope Hilarus was succeeded in 468 by

Simplicius, and in 474 the emperor Leo died, leaving the throne to an infant grandson of the same name, the son of his daughter Ariadne, by an Isaurian officer Zeno, who reigned at first as the guardian of his son, and a few months afterwards came by that son's death to sole power as emperor. The worst character is given to Zeno by the national historians. His conduct was so vile, and his government so discredited by irruptions of the Huns on the Danube, and of Saracens in Mesopotamia, that his wife's stepmother Verina, the widow of Leo I., conspired against him, and was able to set her brother Basiliscus on the throne. Zeno took flight; Basiliscus was proclaimed emperor. He declared himself openly against the Catholic faith in favour of the Eutycheans. But Basiliscus was, if possible, viler than Zeno, and after twenty months Zeno was brought back. The usurper's short rule lasted from October, 475, to June, 477; exactly, therefore, at the time when Odoacer put an end to the western empire. It was upon Zeno's recovery of the throne that he received back from the Roman senate the sovereign insignia, and conferred the title of Roman Patricius on Odoacer. In the following years Zeno had much to do with Theodorich. He gave up to him part of Dacia and Mœsia, and finally he made, in 484, the king of the Ostrogoths Roman consul, as a reward for the services to the Roman emperor. But, afterwards, Theodorich ravaged Zeno's empire up to the walls of Constantinople, and was bought off by a commission to march into Italy and to dethrone Odoacer. Zeno continued an inglorious and unhappy reign, full of murders, deceits, and crimes of every sort, for fourteen years after his restoration, and died in 491.

Let us now pass to the ecclesiastical policy of Zeno's reign.

The succession to the see of Constantinople requires to be considered in apposition with that of the see of Rome. The attempt of Anatolius had been broken by St. Leo, who also outlived him by three years, for Anatolius died in 458, a year after the emperor Leo had succeeded Marcian; and his crowning of Leo is recorded as the first instance of that ceremony being exercised. At his death Gennadius was appointed, who sat to the year 471. He is commended by all writers for his admirable conduct. St. Leo[1] had sent bishops to Constantinople to ask the emperor that he would bring to punishment Timotheus the Cat, who, being schismatical, excommunicated, and Eutychean, had nevertheless got possession of the see of Alexandria. He was endeavouring, after the death of the legitimate bishop, Proterius, who had succeeded the deposed Dioscorus, to ruin the Catholic faith throughout Egypt. All the bishops of the East, whom the emperor consulted, pronounced against this Timotheus. But he was supported by Aspar, who had given Leo the empire. Nevertheless, Gennadius joined his efforts with those of the Pope, and Timotheus Ailouros was banished from Alexandria to Gangra. Another Timotheus Solofaciolus, approved by Pope Leo, was made bishop of Alexandria.

[1] Tillemont, xvi. 68.

At the end of 471, Acacius succeeded Gennadius in the see of the capital. At the time he was well known, having been for many years superior of the orphans' hospital, where he had gained the affection of every-one. He is said to have been made bishop by the influence of Zeno, who was then the emperor's son-in-law. He immediately rose high in the opinion of Leo, who consulted him on private and public affairs before anyone else. He placed him in the senate, the first time that the bishop had sat there. Acacius is said to have used his influence with Leo to soften a severe temper, to restore many persons to his favour, to obtain the recal of many from banishment. He took special care of the churches, and of the clergy serving them, and they in return put his portrait everywhere. Acacius was considered an excellent bishop when Basiliscus rose against Zeno.

In all this contest Acacius took part against the attempt which Basiliscus made to overthrow the faith of the Church. He had issued a document termed the Encyclikon or Circular, in which for the first time in the history of the Church an emperor had assumed the right, as emperor, to lay down the terms of the faith. In this act there is not so much to be considered the mixture of truth and falsehood in the document issued as the authority which he claimed to set up a standard of doctrine. But he could not induce Acacius to put his signature to it. Five hundred Greek bishops, it is true, were found to do so, but Acacius was not one of them. Basiliscus fell, Zeno was restored, and Acacius came out of the struggles between them with increased renown.

Zeno's restoration was considered at the time a victory of the Catholic cause. Basiliscus in his short dominion of twenty months had formally recalled from exile the notorious heretic Timotheus Ailouros, and put him in the patriarchal see of Alexandria, as likewise Peter the Fuller in the see of Antioch. This Timotheus had moved Basiliscus to the strong act of despotically overriding the faith by issuing an edict upon doctrine. Basiliscus had been obliged, by the opposition of the monks at Constantinople, and that of Acacius, and the fear of the returning Zeno, to withdraw this document. The usurper had to fly for refuge to sanctuary, but Acacius did not shield him as St. Chrysostom had shielded Eutropius. He came forth under solemn promise from Zeno that his blood should not be shed, and was carried with wife and children to Cappadocia, where all were starved to death.

In all this matter Acacius had gained great credit as defender of the Council of Chalcedon. He had himself referred for help to Simplicius in the Apostolic See. Zeno upon his return to power had entered into closer connection with the Roman chair. He had sent the Pope a blameless confession of faith, promising to maintain the Council of Chalcedon. Simplicius, on the 8th October, 477, had congratulated him on his return. In this letter he reminds Zeno of the acts of his predecessors, Marcian and Leo: that he owed gratitude to God for bringing him back. "He has restored their empire to you : do you show Him their service. And as the words which I lately addressed, under the instruction of the blessed Apostle Peter, were rejected by those who were about to fall (*i.e.*, Basiliscus), I pray that by God's favour they may profit those who shall stand (*i.e.*, Zeno). I receive the letters sent by your clemency, as an immense pledge of your devotion. I breathe again joyously, and do not doubt that you will do even more in religion than I desire. But mindful of my office, I dwell the more on this matter, because out of regard alike for your empire and your salvation I ardently wish that you should abide in that cause on which alone depends the stability of present government and the gaining future glory. I beg above all things that you should deliver the Church of Alexandria from the heretical intruder, and restore it to the Catholic and legitimate bishop, and also restore the several ejected bishops to their sees, that as you have delivered your commonwealth from the domination of a tyrant, so you may save the Church of God everywhere from the robbery and contamination of heretics. Do not allow that to prevail which the iniquity of the times and a spirit as rebellious against God as against your empire has stirred up, but rather what so many great pontiffs, and with them the consent of the universal Church, has decreed. Give full legal vigour to the decrees of the Council of Chalcedon, or those which my predecessor Leo, of blessed memory, has with apostolic learning laid down. That is, as you have found it, the Catholic faith, which has put down the mighty from their seat, and exalted the humble."[1]

To appreciate this letter, it must be borne in mind that it was written by Pope Simplicius a year after the western empire was extinguished; that the writer had seen nine western emperors deposed, and most of them murdered, in twenty-one years; that it was addressed to the eastern and now only Roman emperor; and that the writer was living under the absolute rule of the *condottiere* chief who had succeeded Ricimer, and is called by Pope Gelasius a few years afterwards "Odoacer, barbarian and heretic".[2]

The whole East was disturbed at this time by the condition of the great patriarchal sees of Alexandria and Antioch. The Eutychean party was perpetually trying for the mastery. At Alexandria, Proterius, who succeeded Dioscorus when he was deposed at the Council of Chalcedon, had been murdered in 458. The utmost efforts of Pope Leo and the emperor Leo were needed to maintain his legitimate successor Timotheus Solofaciolos, against whom a rival of the same name, Timotheus Ailouros, had been set up by the Eutychean party, which was far the most numerous. It was on the death of this patriarch, Timotheus Solofaciolos, in 482, that the clergy and many bishops had chosen John Talaia as his successor. John Talaia had announced his election to the Pope in order to be acknowledged by him; also, as was customary, to the patriarch of Antioch; but had sent his synodal letter by some

---

[1] Simplicii, *Ep.* viii.; Photius, i. 115.   [2] Pope Gelasius, 13th letter.

indirect manner to Acacius, who thus received the notice by public report, rather than in the official way. But in the four years which had elapsed since the restoration of Zeno, Acacius had acquired great influence over him. Zeno had published a decree in which, " out of regard to our royal city," he assured to that " Church, the mother of our piety and the see of all orthodox Christians, the privileges and honours over the consecration of bishops which, before our government, or during it, it is recognised to possess," in which he named Acacius, " the most blessed patriarch, father of our piety". Acacius had made his maintenance of the Council of Chalcedon go step by step with his claim to exercise patriarchal rights over the great see of Ephesus. This had led to fresh reclamations from the Pope. Acacius had gone ever forwards, and seemed, by the favour of Zeno, to be reaching complete subjection of the eastern patriarchates to the see of Constantinople. Incensed at what he considered the slight offered to him by John Talaia, he took up, with the utmost keenness against him, the cause of a rival, Peter the Stammerer, who had been elected by the Eutychean party. He worked upon the emperor's mind in favour of the Monophysite pretender. Peter the Stammerer himself came to Constantinople, and urged to Zeno that the utmost confusion and disorder might be feared in Egypt if the powerful and numerous opponents of the Council of Chalcedon had an unacceptable patriarch put upon them. At the same time, he proposed a compromise which would unite all parties and prevent the breaking up of the eastern Church. Acacius, a few years before, had denounced to Pope Simplicius himself this Peter the Stammerer as an adulterer, robber, and son of darkness. He now entirely embraced this plan, and not only won the emperor to Peter's side for the patriarchate, but induced Zeno to publish a doctrinal decree. This was to express what was common to all confessions of faith down to the Council of Chalcedon, to avoid the expressions used in controversy, and entirely to set aside the Council of Chalcedon. In 482 appeared this Formulary of Union, or Henotikon, drawn up, it was supposed, by Acacius himself, addressed to the clergy and people of Alexandria. It was first subscribed by Acacius, as patriarch of Constantinople, then by Peter the Stammerer, acknowledged for this purpose as patriarch of Alexandria; then by Peter the Fuller, as patriarch of Antioch; by Martyrius of Jerusalem, and by other bishops, but by no means all. Zeno used the imperial power to expel those who would not sign it.

As Peter the Stammerer had gone to the emperor to get his election approved and supported by Zeno and Acacius, so John Talaia had solicited Pope Simplicius to confirm his election. This the Pope had been on the point of confirming, when he received a letter from the emperor accusing John Talaia, and urging the appointment of Peter the Stammerer. Acacius had not hesitated to absolve him, and admit him to his communion, and strove by every effort of deceit and force to induce the eastern bishops to accept him. The last letter we have of the Pope, dated November 6, 482, strongly censures Acacius for communicating nothing to him concerning the Church of Alexandria, and for not instructing the emperor in such a way that peace might be restored by him.

On March 2, 483, Pope Simplicius died, and was succeeded by Pope Felix. John Talaia had come in person to Rome to lay his accusation against Acacius. Also the orthodox monks at Constantinople, and eastern bishops expelled for not signing the Henotikon, begged for the Pope's assistance, and denounced Acacius as the author of all the trouble. Amongst these expelled bishops who appealed to Rome were bishops of Chalcedon, Samosata, Mopsuestia, Constantina, Hemeria, Theodosiopolis.

The Pope called a council, in which he considered the complaint now brought before him by John Talaia, as a hundred and forty years before St. Athanasius had carried his complaint to Pope Julius. It was resolved to support the ejected bishops, to maintain the Council of Chalcedon, and to request from the emperor the expulsion of Peter the Stammerer, who was usurping the see of Alexandria. For this purpose the Pope commissioned two bishops, Vitalis and Misenus, to go as his legates to the emperor. They were to invite Acacius to attend a council at Rome, and to answer therein the complaint brought against him by the elected patriarch of Alexandria.

The legates carried a letter[1] from Pope Felix to the emperor, in which, according to custom, the Pope informed him of his election. He observed that, for a long time, the see of the blessed Apostle had been

---

[1] Mansi, vii. 1032-6; Jaffé, 359.

expecting an answer to the letters sent by his predecessor of blessed memory, "especially inasmuch as it had bound your majesty, with tremendous vows, not to allow the see of the evangelist St. Mark to be separated from the teaching or the communion of his master. . . . Again, therefore, the reverend confession of the Apostle Peter, with a mother's voice, renews its instance. It ceases not with confidence to call upon you as its son. It cries: O Christian prince, why do you allow me to be interrupted in that course of charity which binds together the universal Church? Why, in my person, do you break up the consent of the whole world? I beseech you, my son, suffer not that tunic of the Lord woven from the top throughout, by which is signified, as the Holy Spirit rules the whole body, that the Church of Christ should be one and individual—suffer it not to be broken. They who crucified our Saviour left it untouched. Do not let it be rent in your times. My faith it is which the Lord Himself declared should alone be one, never to be conquered by any assault: He who promised that the gates of hell should never prevail over the Church founded on my confession. This Church it was which restored you to the imperial dignity, deprived its impugners of their power, and opened to you the path of victory in defending it.[1]

"Look at me, his successor, however humble, as if the Apostle were present. Look deeper into those ways which concern the reverence due to God and the condition of man; and be not ungrateful to the

[1] Mansi, vii. 1028; Jaffé, 360.

Author of your present prosperity. In you alone survives the name of emperor. Do not grudge us the saving you. Do not diminish our confidence in praying for you. Look back on your august predecessors Marcian and Leo, and the faith of so many princes, you, who are their lawful heir. Once more, look back on your own engagements, and the words which, on your return to power, you addressed to my predecessor. The defence of the Council of Chalcedon is expressed in the whole series." And he ends: "What I could not put in my letter I have entrusted my brethren and legates to explain. I beseech you to listen, as well for the preservation of Catholic truth as for the safety of your own empire."

To Acacius also the legates carried a letter of the Pope, which he opened by announcing that he had succeeded to the office of Pope Simplicius, and was forthwith involved in those many cares which the voice of the Supreme Pastor had imposed upon St. Peter, and which kept him watchfully occupied with a rule which extended over all the peoples of the earth. At that moment his greatest anxiety, as it had been that of his predecessor, was for the city of Alexandria, and for the faith of the whole East. And he went on to reproach Acacius for not duly informing him of what was passing, for not defending the Council of Chalcedon, and not using his influence with the emperor in its defence: "Brother, do not let us despair that the word of our Saviour will be true; He promised that He would never be wanting to His Church to the end of the world; that it should never be overcome by the gates of hell; that all which was bound on earth by sentence of apostolic doctrine should not be loosed in heaven. Nor let us think that either the judgment of Peter or the authority of the universal Church, by whatever dangers it be surrounded, will ever lose the weight of its force. The more it dreads being weakened by worldly prosperity, the more, divinely instructed, it grows under adversity. To let the perverse go on in their way, when you can stop them, is indeed to encourage them. He who, evidently, ceases to obstruct a wicked deed, does not escape the suspicion of complicity. If, when you see hostility arising against the Council of Chalcedon, you do nothing, believe me, I know not how you can maintain that you belong to the whole Church."

As soon as the two legates arrived at the Dardanelles, they were arrested, by order of Zeno and Acacius, put in prison, their papers and letters taken from them. They were menaced with death if they did not accept the communion of Acacius and of Peter the Stammerer. Then they were seduced with presents, and deceived with false promises that Acacius would submit the whole affair to the Pope. They resisted at first, but yielded in the end, and, passing beyond their commission, gave judgment in favour of Peter the Stammerer. They had broken all the instructions of the Pope, and carried back letters from Zeno and Acacius to him, full of extravagant praises of Peter the Stammerer. His former deposition and condemnation were entirely put aside. On the other hand, the character of John Talaia was bitterly impugned. The emperor asserted that he had treated Church matters with the utmost moderation, and guided himself entirely by the advice of the patriarch Acacius.

In fact, Acacius was the spiritual superior of the whole eastern empire, and appeared not to trouble himself any more about the Roman See. He made no pretence to give any satisfaction for what he had done. Before he had been the champion of orthodoxy, now he had become in league with heretics. But he lost all remaining confidence among Catholics. The zealous monks of his own city withdrew from his communion, and sent one of themselves, Symeon, to Rome to inform the Pope of all that had happened, and disclose the faithless behaviour of his legates.[1]

In another letter the Pope had cited Acacius to appear at Rome to meet the accusation brought against him by John Talaia, the patriarch of Alexandria. Acacius took no notice of this citation, nor of the complaint brought against him.

Thereupon, the Pope, in a council of seventy-seven bishops, held at Rome the 28th July, 484, made inquiry into all this transaction. He annulled the judgment on Peter the Stammerer, passed without his authority by his legates, deprived them of their offices, and of communion. He renewed the condemnation of Peter the Stammerer. He had in the interval admonished Acacius again, without result. He now issued the decree of deposition upon him. It runs in the following words:

[1] Photius, i. 123, translated.

"You are[1] guilty of many transgressions; have often treated with insult the venerable Nicene Council; have unrightfully claimed jurisdiction over provinces not belonging to you. In the case of intruding heretics, ordained likewise by heretics, whom you had yourself condemned, and whose condemnation you had urged upon the Apostolic See, you not only received them to your communion, but even set them over other Churches, which was not, even in the case of Catholics, allowable; or have even given them higher rank undeservedly. John is an instance of this. When he was not accepted by the Catholics at Apamea, and had been driven away from Antioch, you set him over the Tyrians. Humerius also, having been degraded from the diaconite and deprived of the Christian name, you advanced to the priesthood. And as if these seemed to you minor offences, in the boldness of your pride you assaulted the truth itself of apostolic doctrine. That Peter, whose condemnation by my predecessor of holy memory you had yourself recorded, as the subjoined proofs show, you suffered by your connivance again to invade the see of the blessed evangelist Mark, to drive out orthodox bishops and clergy, and ordain, no doubt, such as himself, to expel one who was there regularly established, and hold the Church captive. Nay, his person was so agreeable to you, and his ministers so acceptable, that you have been found to persecute a large number of orthodox bishops and clergy, who now come to Constantinople, and to encourage his legates.

[1] Mansi, vii. 1065; Baronius (anno 484), 17; Jaffé, 364.

You put upon Misenus and Vitalis to find excuse for one who was anathematising the decrees of the Council of Chalcedon, and violating the tomb of Timotheus of holy memory, as sure information has been given us. You have not ceased to praise and exalt him so as to boast that the very condemnation you had yourself recorded was untrue. You went even further in the defence of a perverse man. They who were late bishops, but are now deprived of their rank and of communion, Vitalis and Misenus, men whom we had specially sent for his expulsion, you suffered to be deprived of their papers and imprisoned; you dragged them out thence to a procession which you were having with heretics, as they confessed; in contempt of their legatine quality, which even the law of nations would protect, you drew them on to the communion of heretics, and yourself; you corrupted them with bribes; and, with injury to the blessed Apostle Peter, from whose see they went forth, you caused them not only to return with labour lost, but with the overthrow of all their instructions. In deceiving them, your wickedness was shown. As to the memorial of my brother and fellow-bishop John (Talaia), who brought the heaviest charges against you, by not venturing to give an answer in the Apostolic See, according to the canons, you have established his allegations. Likewise, you considered unworthy of your sight our most faithful defender Felix, whom a necessity caused to come afterwards. You also showed by your letters that known heretics were communicating with you. For what else are they who, after the death of Timotheus of holy memory, go back to his church under Peter the Stammerer, or, having been Catholics, have given themselves up to this Peter, but such as Peter himself was judged to be by the whole Church, and by yourself? Therefore, by this present sentence have with those whom you willingly embrace your portion, which we send to you by the defender of your own church, being deprived of sacerdotal honour and Catholic communion, and severed from the number of the faithful. Know that the name and office of the sacerdotal ministry is taken from you. You are condemned by the judgment of the Holy Ghost[1] and apostolic authority, and never to be released from the bonds of anathema.

"Cœlius Felix, bishop of the holy Catholic Church of the city of Rome. On the 28th July, in the consulship of the most honourable Venantius."

This was a synodal letter,[2] signed by sixty-seven bishops, as well as the Pope. But the copy of the decree against Acacius sent to Constantinople was signed by the Pope alone, partly according to ancient custom, partly in order with greater security to transmit it to the eastern capital. Had this copy been signed by the bishops also, ruling practice would have required it to be carried over by at least two bishops, which then appeared very dangerous. A Roman synod of forty-

---

[1] It is to be observed that the Pope calls his judgment the Judgment of the Holy Ghost, just as Pope Clement I. did in the first recorded judgment. See his letter, secs. 58, 59, 63, quoted in *Church and State*, 198-199.

[2] Photius, i. 124.

three[1] bishops, in the following year, 485, wrote to the clergy of Constantinople: "If snares had not been set for the orthodox by land and sea, many of us might have come with the sentence of Acacius. But now, being assembled on the cause of the church of Antioch at St. Peter's, we make a point of declaring to you the custom which has always prevailed among us. As often as bishops[2] meet in Italy on ecclesiastical matters, especially when they touch the faith, the custom is maintained that the successor of those who preside in the Apostolic See, as representing all the bishops of the whole of Italy, according to the care of all churches which lies upon him, appoints all things, being the head of all, as the Lord said to Peter, 'Thou art Peter,' &c. The three hundred and eighteen holy fathers assembled at Nicæa acted in obedience to this word, and left the confirmation and authority of what they treated to the holy Roman Church; both of which things all successions to our own time by the grace of Christ maintain. What, therefore, the holy council assembled at St. Peter's decreed, and the most blessed Felix, our Head, Pope, and Archbishop, ratified, that is sent to you by Tutus, defensor of the Church."

Three days after the sentence on Acacius, Pope Felix wrote to the emperor Zeno.[3] He reminded him that, in violation of reverence to God, an embassy to the Holy See had been taken captive, its papers taken away; it had been dragged out of prison to communicate with

---

[1] Mansi, vii. 1139; Baronius (anno 484), 26, 27.
[2] Domini sacerdotes.   [3] Jaffé, 365; Mansi, vii. 1065.

the officers of the very heretic against whom it had been sent. Since even barbarous nations, who knew not God, allowed to embassies for the transaction of human affairs a sacred liberty, how much more should that liberty be preserved sacred, especially in divine things, by a Roman emperor and Christian prince? Putting aside the embassy, which even in the case of the Apostle Peter was disregarded, be assured at least by these letters that the see of the Apostle Peter has never granted communion, and will never grant it, to that Alexandrian Peter long ago justly condemned, and again by synodal decree suppressed. But as you have not regarded the words of exhortation I addressed to you, I leave it to your choice to select which you will have, the communion of the blessed Apostle Peter or that of the Alexandrian Peter. You will know by the letters of this man's abettor, Acacius, to my predecessor of holy memory, copies of which I enclose, how even in your own judgment he was condemned. But this Acacius, who has committed many atrocities against the ancient rules, and has come to praise one whom he affirmed to be condemned, and whose condemnation he obtained from the Apostolic See, has been severed from apostolic communion. But I believe that your piety, which prefers to comply even with its own laws rather than to resist them, and which knows that the supreme rule of things human is given to you on condition of admitting that things divine are allotted to dispensers divinely assigned, I believe that it will be undoubtedly of service to you if you permit the

Catholic Church in the time of your principate to use its own laws, nor allow anyone to stand in the way of its liberty, which has restored to you the imperial power. For it is certain that this will bring safety to your affairs, if in God's cause, and according to His appointment, you study to subdue the royal will and not to prefer it to the bishops of Christ, and rather to learn holy things by them than to teach them; to follow the form traced out by the Church, not after human fashion to impose rules on it, nor wish to dominate the commands of that power to whom it is God's will that your clemency should devoutly submit, lest, if the measure of the divine disposition be overpast, it may end in the disgrace of the disponent. And from this time I absolve my conscience as to all these things, who have to plead my cause before Christ's tribunal. It will be well for you more and more to reflect that both in the present state of things we are under the divine examination, and that after this life's course we shall according to it come before the divine judgment."

St. Gregory the Great, writing his *Dialogues*[1] about one hundred and ten years after this letter, informs us that the writer of it was his great-grandfather, and speaks of his appearing in a vision to his aunt Tarsilla and showing her the habitation of everlasting light. At the time of writing it, Pope Felix was living under the domination of the Arian Herule Odoacer. The great Church of Africa was suffering the most terrible of persecutions under the Arian

[1] iv. 16.

Vandal Hunneric, the son of his father Genseric. Arian Visigoth rulers were in possession of Spain and France, of whom Euric, as we have seen, was described rather as the chief of a sect than the sovereign of a people. In all the West not a yard of territory was under rule of a Catholic sovereign. And he whom the Pope addressed, with the dignity of the Apostolic See in its reverence for the power which is a delegation of God, as Roman emperor and Christian prince, was in his private life scandalous, in all his public rule shifty and tyrannical, and in belief, if he had any, an Eutychean heretic. It may be added, as a fact of history, that the emperor went before the divine judgment sooner than the Pope; that during the seven years which intervened between the letter and his death he utterly disregarded all that the Pope had done and said. He suffered, or rather made the bishop of Constantinople to be the ruler of the eastern Church; he maintained heretics in the sees of Alexandria and Antioch. After this he died in 491, and the last fact recorded of him is that the empress Ariadne, the daughter of Leo I., who had brought him the empire with her hand, when he fell into an epileptic fit and was supposed to be dead, had him buried at once, and placed guards around his tomb, who were forbidden to allow any approach to it. When the imperial vault was afterwards entered, Zeno was found to have torn his arm with his teeth. The empress widow, forty days after the death of Zeno, conferred her hand, and with it the empire a second time, upon Anastasius, who had been up to that time a sort of gentleman usher[1] in the imperial service. Anastasius ruled the eastern empire twenty-seven years, from 491 to 518.

The Pope further sought by a letter[2] to the clergy and people of Constantinople to remove the scandal caused by the weakness of his legates, and to explain the grounds upon which he had deposed Acacius. "Though we know the zeal of your faith, yet we warn all who desire to share in the Catholic faith to abstain from communion with him, lest, which God forbid, they fall into like penalty."

Acacius did not receive the papal judgment against him, but sought to suppress it. A monk ventured to attach to his mantle as he went to Mass the sentence of excommunication. It cost him his life, and brought heavy persecutions on his brethren. Acacius met the Pope with open defiance, and removed his name from the diptychs.[3] He rested on the emperor Zeno's support, who did everything at his bidding. Every arm of deceit and of violence he used equally. The monks, called, from their never intermitted worship, the Sleepless, in close connection with Rome, suffered severely. So Acacius passed the remaining five years of his life, dying in the autumn of 489.

---

[1] Silentiarius, in the Greek court, officers who kept silence in the emperor's presence.

[2] *Ep.* x.; Mansi, vii. 1067.

[3] "The recital of a name in the diptychs was a formal declaration of Church fellowship, or even a sort of canonisation and invocation. It was contrary to all Church principles to permit in them the name of anyone condemned by the Church."—*Life of Photius*, i. 133, by Card. Hergenröther.

His excommunication by the Pope caused a schism between the East and West which lasted thirty-five years, from 484 to 519. He met that supreme act of authority by the counter act of removing the Pope's name from the diptychs. This invites us to consider the position which he assumed.

From the year 482 (that is, four years after Zeno had recovered the empire), Acacius appears in possession of full influence over the emperor. The position of the bishop at Constantinople was, in itself, one of immense dignity. He was undoubtedly the second person in the imperial city, surrounded with a pomp and deference only yielding to that accorded to the emperor, but in some respects superior to it. He was regarded as sacrosanct: all the respect which the Church received in the minds of the good was centred in his person. And as he had risen to all this dignity in virtue of Constantinople being the capital, there was a special connection between the capital and its bishop, which led it to sympathise with every accession of power which he received. There can be no doubt that the right acquired by that bishop over the great sees of Ephesus, Cæsarea in Pontus, and Heraclea in Thrace was extremely popular at Constantinople; and that when he proceeded further to show his hand over the patriarchate of Antioch—as, for instance, in nominating one of its archbishops at Tyre, as the Pope reproached him—the capital was still better pleased. Most of all when, breaking through all the regulations which the Nicene Council had consecrated by its approval,—

which, however, it had not created, but found in immemorial subsistence,—he ventured to ordain at Constantinople a patriarch of Antioch. Thus Stephen II., patriarch of Antioch, had been murdered in 479 by the fanatical Monophysites, in the baptistry of the Barlaam Church, and his mangled body thrown into the Orontes. The incensed emperor punished the criminals, and charged his patriarch Acacius to consecrate a new bishop for Antioch. Acacius seized the favourable opportunity, after the example of Anatolius, to advance himself, and appointed Stephen III. Emperor and patriarch both applied to Pope Simplicius to excuse this violation of the rights of the Syrian bishops, alleging the pressure of circumstances, and promising that the example should not occur again. Simplicius, so entreated, excused the fault, recognised the patriarch of Antioch—though he had been consecrated in Constantinople by its bishop—but insisted that such a violation of the canons should not be repeated. Presently Stephen III. died, upon which Acacius committed the same fault anew, and in 482 consecrated Calendion patriarch of Antioch. Calendion brought back from Macedonia the relics of his great and persecuted predecessor, St. Eustathius; but presently Zeno and Acacius displaced Calendion. Acacius was using the power which he possessed over the emperor to advance his own credit in the appointment of patriarchs, and to establish two notorious heretics—Peter the Fuller at Antioch, and Peter the Stammerer at Alexandria. All this meant that the bishop of Constantinople's hand was to be over the East, as the bishop of Rome's hand was over the West. Then, ever since the Council of Chalcedon, the two great eastern patriarchates had been torn to pieces by the conflicts of parties. The Eutychean heresy fought a desperate battle for mastery. As to Antioch, from the time that Eusebius of Nicomedia had brought about the deposition of St. Eustathius, preparatory to that of Athanasius in 330, the great patriarchate of the East had been declining from the unrivalled position which it had held. As to Alexandria, from the time that the 150 fathers at Constantinople, in 381, had attempted to make Constantinople the second see, because it was Nova Roma, the see of St. Mark bore a grudge against the upstart which sought to degrade it. In spite of the unequalled renown of its two great patriarchs, St. Athanasius and St. Cyril, it was sinking. And now heresy, schism, and imperial favour seemed to have joined together to exhibit Acacius as not only the first patriarch of the East, but as exercising jurisdiction even within their bounds, and as nominating those who succeeded to their thrones. All which would only tend to increase the power and popularity of the bishop of Constantinople in his own see.

Acacius had now been eleven years bishop. He had gained at once the emperor Leo; he had appeared to defend the Council of Chalcedon when Basiliscus attacked it; he had further gained mastery over Zeno; but, more than all this, he had seen Rome sink into what to eastern eyes must have seemed an abyss. St. Leo had compelled Anatolius to give up the canons he so much prized;

since then northern barbarians had twice sacked Rome, and Ricimer's most cruel host of adventurers had reaped whatever the Vandal Genseric had left. If there was a degradation yet to be endured it would be that a Herule soldier of fortune should compel a Roman senate to send back the robes of empire to Constantinople, and be content to live under a Patricius, sprung from one of the innumerable Teuton hordes, and sanctioned by the emperor of the East; and Acacius would not forget that in the councils of that emperor he was himself chief.

If New Rome held the second rank because the Fathers gave the first rank to Old Rome, in that it was the capital, what was the position of New Rome and its bishop when Old Rome had ceased in fact to be a capital at all? At that moment—thirty years after St. Leo had confirmed the greatest of eastern councils and been greeted by it as the head of the Christian faith— the Rome in which he sat had been reduced to a mere municipal rank, and its bishop, with all its people, lived under what was simply a military government commanded by a foreign adventurer. Odoacer at Ravenna was master of the lives and liberties of the Romans, including the Pope.

Acacius had had this spectacle for some years before him, when Pope Felix, succeeding Pope Simplicius, called him to account for entirely reversing the conduct which he had pursued at the time when Basiliscus had usurped the empire. Then he defended the Council of Chalcedon and its doctrine; then he denounced to the Pope Peter the Stammerer as a heretic and a man of bad life, and had called for his condemnation and obtained it. He had now taken upon himself not even to ask from the Pope this man's absolution, but to absolve himself the very heretic he had caused to be condemned, and to put him into the see of Alexandria, with the rejection of the bishop legitimately elected, and approved at Rome, and to compose for the emperor a doctrinal decree, which he subscribed himself first as the first of the patriarchs, and was compelling all other bishops to sign under pain of deprivation; when, behold, St. Leo's third successor called him to account in exactly the same terms as St. Leo would have used, and required him to meet at Rome the accusation brought against him by John Talaia, a duly elected patriarch of Alexandria, just as St. Julius, a hundred and forty years before, had invited the accusing bishops at Antioch to meet St. Athanasius before his tribunal. He who resided in a state only second to the emperor in the real capital of the empire to go to a city living in durance under the northern barbarians, and submit to the judgment of one whose own tribunal was in captivity to such masters!

But, on the other hand, Pope Felix spoke to the emperor as none but popes have ever spoken. He called him his son, but he required from him filial obedience. Above all he spoke in one character, and in one alone—as the heir of that St. Peter whom the voice of the Lord had set over His Church; he spoke from Rome, not because it was or had been capital of the empire, but because it was St. Peter's See, and precisely because he succeeded St. Peter in his apostolate.

The respective action, therefore, of Pope Felix on one side, and of Acacius on the other, brought to an issue the most absolute of contradictions. The Pope claimed obedience, as a superior, from Acacius. When that obedience was refused, he exerted his authority as superior, and degraded Acacius both from his rank as bishop, and from Christian communion. And a special token of that sentence was to order his name to be removed from the diptychs, and to enjoin the people of his own diocese to hold no communion with him, on pain of incurring a like penalty with him. Acacius answered by practically denying the Pope's authority to do any such act. He asserted himself to be his equal by removing the Pope's name from the diptychs. There could be no more striking denial of any such authority as the claim to inherit Peter's universal pastorship, than to treat the Pope himself as, in virtue of that pastorship, he had treated Acacius.

Even apart from this, the conduct of Acacius carried with it a double denial of the Pope's authority: a denial that he was the supreme judge of faith; and a denial that he was the supreme maintainer of discipline in its highest manifestation, the order of the hierarchy itself.

He denied that the Pope was the supreme judge of faith, by drawing up a formulary of doctrine, which he induced the emperor to promulgate by imperial decree; and this independently of what doctrine that formulary might contain. Further, he did this by supporting two persons judged to be heretical by the Holy See—Peter the Fuller at Antioch, Peter the Stammerer at Alexandria. He denied that the Pope was the supreme maintainer of discipline, by making the two great sees of the East and South subordinate to himself. As the Pope expressed it in his sentence, he had done "nefarious things against the whole Nicene constitution," of which the Pope was special guardian. In fact, his conduct was an imitation of that pursued in the preceding century by Eusebius of Nicomedia, by Eudoxius, and all their party. It was even carried out to its full completion. The emperor was made the head of the Church, on condition of his leading it through the bishop of Constantinople. Acacius put together the canon of the Council of 381, which said that the bishop of New Rome should hold the second rank in the episcopate, because his city is New Rome, with the canon attempted to be passed at Chalcedon, and cashiered by St. Leo, that the fathers gave its privileges to Old Rome because it was the imperial city. Uniting the two, he constructed the conclusion, that as Old Rome had ceased to be the imperial city, which New Rome had actually become, the privileges of Old Rome had passed to the bishop of New Rome.

This he expressed by removing the name of the Pope from the diptychs in answer to his sentence of degradation and excommunication. As the Pope could not suffer the conduct of Acacius, without ceasing to hold the universal pastorship of St. Peter, so Acacius could not submit to it without admitting that pastorship. He denied it in both its heads of faith and government by his conduct. He embodied that denial unmistakably in removing the Pope's name from the diptychs.

To lay down a parity between the ecclesiastical privileges of the two sees, Rome and Constantinople, because their cities were both capitals, is implicitly to deny altogether the divine origin of ecclesiastical jurisdiction. That is, to deny that the Church is a divine polity at all. The conduct of Acacius was to bring that matter to an issue. The end of it will show whether he was right or wrong.

He lived for five years, from 484 to 489, strong in the emperor's support, who did everything which he suggested. And he had his part as a counsellor, as well as a bishop, in one most important transaction, which took place in this interval. The reign of Zeno was disturbed by perpetual insurrections and perils. In these Theodorick the Goth had been of great service to him, so that in this year, 484, Zeno had made him consul at Rome. But Theodorick afterwards thought that Zeno had treated him very ill. He marched upon Constantinople: Zeno trembled on his throne. Something had to be done. What was done was to turn Theodorick's longing eyes upon the land possessing "the hapless dower of beauty".[1] Zeno commissioned him to turn Odoacer out, and to take his place. In

---

[1] "Cui feo la dote
Dono infelice di bellezza, ond' hai
Funesta dote d'infinite guai."
—*Filicaja*.

489, Theodorick led the great mass of his people into Italy, at the suggestion, and with the warrant of, the man whom Pope Felix had appealed to as his son, the Roman emperor and Christian prince. And so, as an emperor and a bishop of Constantinople, a hundred years before, had led the Gothic nation into the Arian heresy, under the belief that it was the Christian faith, another emperor of Constantinople and another bishop turned that Gothic nation upon the Roman mother and the See of Peter, regardless that they would thereby become temporal subjects of those who were possessed by the "Arian perfidy". Beside Eudoxius and Valens in history stand Acacius and Zeno; and beside Alaric, let loose with his warlike host by the younger sister on the elder in 410, stands Theodorick, commissioned, in 489, with all his people, to occupy permanently the birthplace of Roman empire.

The eastern bishops[1] crouched before the emperor's power and his patriarch's intrigues, who deposed those who were not in his favour, and tyrannised over the greater number, so that many fled to the West. John Talaia himself, the expelled patriarch of Alexandria, received the bishopric of Nola from the Pope, to whom he had appealed. This continued to be the state of things during five years, from 484 to 489, when Acacius died, still under sentence of excommunication. One of the greatest bishops of his time, St. Avitus of Vienna, characterises him with the words, "Rather a timid

---

[1] Photius, i. 128, who quotes Avitus, 3rd letter, and Ennodius, and Gelasius, *Ep.* xiii.

lover than a public asserter of the opinion broached by Eutyches: he praised, indeed, what he had taken from him, but did not venture to preach it to a people still devout, and therefore unpolluted by it". Another equally great bishop, Ennodius of Ticinum—that is, Pavia—says: "He utterly surrendered the glory which he had gained, in combating Basiliscus, of maintaining the truth"; while the next Pope Gelasius charges him with intense pride; the effect of which was to leave to the Church "cause for the peaceful to mourn and the humble to weep".

But all this evil had been wrought by Acacius, and upon his death it remained to be seen how his successor would act. He was succeeded by Fravita,[1] who, so far from maintaining the conduct of Acacius in excluding the name of Pope Felix from the diptychs, wished above all things to obtain the Pope's recognition. He would not even assume the government of his see without first receiving it. It was usual for patriarchs and exarchs to enter on their office immediately after election and consecration, before the recognition of the other patriarchs which they afterwards asked for by sending an embassy with their synodal letter. It seems Fravita would make no use of this right, but besought the Pope's confirmation in a very flattering letter. It would seem also that, by the death of Acacius, the emperor Zeno had been delivered from thraldom, and returned to some sentiment of justice. For he supported the letter of the new patriarch by one himself to the Pope, and it is

[1] Photius, i. 126; Hefele, *C. G.*, ii. 596.

from the Pope's extant answers[1] to these two writings
that we learn some of their contents. To the emperor,
the Pope replies that he knows not how to return sufficient thanks to the divine mercy for having inspired
him with so great a care for religion as to prefer it to
all public affairs, and to consider that the safety of the
commonwealth is involved in it. That, desiring to confirm the unity of the Catholic faith and the peace of the
churches, he should be anxious for the choice of a bishop
who should be remarkable for personal uprightness and,
above all things, for affection to the orthodox truth.
That the Church has received in him such a son, and
that the pontiff, in whose accession he rejoices, has already
given an indication of his rule in referring the beginning
of his dignity to the See of the Apostle Peter. For
the newly-elected pontiff acknowledges in his letter that
Peter is the chief of the Apostles and the Rock of the
Faith: that the keys of the heavenly mysteries have
been entrusted to him, and therefore seeks agreement
with the Pope. Then, after enlarging upon the misdeeds of Acacius, and his rejection of the Council of
Chalcedon, and his absolution of notorious heretics, the
Pope beseeches the emperor to establish peace by giving
up the defence of Acacius. "I do not extort this from
you—as being, however unworthy, the Vicar of Peter—
by the authority of apostolic power; but, as an anxious
father earnestly desiring the prosperity of a son, I im-
plore you. In me, his Vicar, how unworthy soever, the
Apostle Peter speaks; and in him Christ, who suffers

[1] Jaffé, 371, 372; Mansi, vii. 1097; vii. 1100.

not the division of His own Church, beseeches you. Take from between us him who disturbs us: so may Christ, for the preservation of His Church's laws, multiply to you temporal things and bestow eternal."

In his answer to Fravita, Pope Felix expresses the pleasure which his election gives, and the hope that it will bring about the peace of the Church. He takes his synodal letter as addressed to the Apostolic See, "through which, by the gift of Christ, the dignity of all bishops is made of one mass,"[1] as a token of good-will, inasmuch as his own letter confesses the Apostle Peter to be the head of the Apostles, the Rock of the Faith, and the dispenser of the heavenly mystery by the keys entrusted to him. He is the more encouraged because the orthodox monks formed part of the embassy. But when the Pope required a pledge from them that Fravita should renounce reciting the names of Peter the Stammerer and Acacius in the church, they replied that they had no instructions on that head. For this reason the Pope delayed to grant communion to Fravita, and he exhorts him, in the rest of the letter, not to let the misdeeds of Acacius stand in the way of the Church's peace. "Inform us then, as soon as possible, on this, that God may conclude what He has begun, and that, fully reconciled, we may agree together in the structure[2] of the body of Christ."

---

[1] Dum scilicet ad Apostolicam Sedem regulariter destinatur, per quam *largiente Christo omnium solidatur dignitas sacerdotum.* Quod ipsæ dilectionis tuæ literæ Apostolorum summum petramque fidei et cælestis dispensatorem mysterii creditis sibi clavibus beatum Petrum Apostolum confitentur.

[2] In compage corporis Christi **consentire.**

Fravita died before he received the answer of the Pope, having occupied the see of Constantinople only three months, and out of communion with the Pope.

It would seem that the first successor of Acacius as well as the emperor receded both from his act and the position which it involved. They acknowledged in their letters, as we learn from the Pope's recitation of their words, the dignity of the Apostolic See. What they were not willing to do was to give up the person of Acacius. What the subsequent patriarchs, Euphemius and Macedonius, alleged, was that he was so rooted in the minds of the people that they could not venture to condemn him by removing his name from commemoration in the diptychs.

In 490, Euphemius followed in the see of Constantinople. He was devoted to the Council of Chalcedon, and ever honoured in the East as orthodox. He replaced the Pope's name in the diptychs, and renounced communion with Peter the Stammerer, who had again openly anathematised the Council of Chalcedon; only he refused to remove from the diptychs the names of his two predecessors. Pope Felix had written, on the 1st May, 490, to the archimandrite Thalassio,[1] not to enter into communion with the bishop who should succeed Fravita, even if he satisfied these demands respecting Acacius and Peter the Stammerer, unless with the express permission of the Roman See. This condition he maintained, acknowledging Euphemius as orthodox, but not as bishop, because he would not

---

[1] Jaffé, 374; Mansi, vii. 1103.

remove from the diptychs the names of two predecessors who had died outside of communion with the Roman See.

Euphemius had himself subscribed the Henotikon of Zeno, without which the emperor would never have assented to his election; but he confirmed in a synod the Council of Chalcedon. When, in April, 491, Zeno died, and through the favour of his widow, the empress Ariadne, Anastasius obtained the throne in a very disturbed empire, the patriarch long refused to set the crown on his head, because he suspected him to favour the Eutychean heresy. The empress and the senate besought him in vain. He only consented when Anastasius gave him a written promise to accept the decrees of Chalcedon as the rule of faith, and to permit no innovation in Church matters. On this condition he was crowned: but emperor and patriarch continued at variance. The emperor tried to escape from his promise in order to maintain Zeno's Henotikon, which he thought the best policy among the many factions of the East. Euphemius was in the most unhappy position with the monks, who would not acknowledge him because he was out of communion with the Pope on account of Acacius.

Pope Felix, having all but completed nine years of a pontificate, in which he showed the greatest fortitude in the midst of the severest temporal abandonment, died in February, 492. Italy then had been torn to pieces for three years by the conflict between Odoacer and Theodorick. Gondebald, king of the Burgundians, had cruelly ravaged Liguria. Then it was that bishops began to build fortresses for the defence of their peoples. The Church of Africa was in the utmost straits under the cruelty of Hunneric. Pope Gelasius succeeded on the 1st March, 492. His pontificate lasted four years and eight months; during the whole course of which his extant letters show that he was no less exposed to temporal abandonment than Felix, and no less courageous in maintaining the pastorship of Peter.

But the death of the emperor Zeno in 491, and the death of Pope Felix III. ten months afterwards, in 492, require us to make a short retrospect of the temporal condition of empire and Church at this time. Zeno, receiving the empire at the death of his young son by Ariadne, Leo II., in 474, had reigned seventeen years, if we comprise therein the twenty months during which the throne was occupied by the insurgent Basiliscus from 475 to 477, precisely at the moment when Odoacer terminated the western empire. Zeno, recovering the throne in 477, had acted as a Catholic during about four years. Pope Simplicius had warmly congratulated him on the recovery of the empire on the 8th October of that year. In 478, the Pope had thanked Acacius for informing him that the right patriarch, Timotheus Solofaciolus, had been restored at Alexandria. But from 482 all is altered. The chronicle of Zeno's reign becomes a catalogue of misfortunes. The publication of his Formulary of Union is a gross attack upon the spiritual independence of the Church. He imposes it upon the eastern bishops on pain of expulsion. He puts open heretics into the sees of Alexandria and Antioch. All this is done under the advice and instigation of Acacius, who is the real author of the Henotikon, and who completes his acts by open defiance of Pope Felix. When Zeno died he left the empire a prey to every misery. In Italy, Herules and Ostrogoths were desperately contending for the possession of the country. Barbarians beyond the Danube incessantly threatened the north-eastern frontiers. There was no truce with them but at the cost of incessant payments and every sort of degradation. Egypt and Syria were torn to pieces by the Eutychean heresy. The infamous surrender of Italy to Theodorick in 488 has been touched upon. By that the support which the Ostrogothic king had given to keep Zeno on a tottering throne, followed by the terror which his discontent had caused at Constantinople, purchased from the Roman emperor himself the sacrifice of Rome and all the land from the Alps to the sea. Such was the man with whom the Popes Simplicius and Felix had to deal. To him it was that, from a Rome which drew its breath under an Arian Herule, the commander of adventurers who sold their swords for hire, these Popes wrote those letters full of Christian charity and apostolic liberty which have been quoted.

When Zeno died in 491, he was attended to the grave by the contempt of his own wife and the malediction of the people, whom his cruelty, debauchery, and perfidy had alienated. I take from an

ancient Greek document[1] a note of what followed.
"When Zeno died, Anastasius succeeded to his wife
and the empire; and he assembled an heretical council
in Constantinople on account of the holy Council of
Chalcedon, in which, by subjecting Euphemius to numberless calumnies, he banished him beyond Armenia,
and put in the see the most blessed Macedonius.
Macedonius called an upright council, and expressly
ratified the decrees of faith passed at Chalcedon; but
through fear of Anastasius he passed over in silence
the Henotikon of Zeno." "When now Peter the Fuller
was cast out of Antioch, Palladius succeeded to the see.
And when he died Flavian accepted the Henotikon of
Zeno; and he expressly confirmed the three holy
Ecumenical Councils, but to please the emperor he
passed over in silence that of Chalcedon. Now the
emperor Anastasius sent order by the tribune Eutropius
to Flavian and Elias of Jerusalem to hold a council in
Sidon, and to anathematise the holy Council of Chalcedon. But Elias dismissed this without effect; for
which the emperor was very indignant with the
patriarchs. But when Flavian returned to Antioch,
certain apostate monks, vehement partisans of the folly

---

[1] The "libellus synodicus," says Hefele, *C. G.*, i. 70, "auch synodicon genannt, enthält kurze Nachrichten über 158 Concilien der 9 ersten Jahrhunderte, und reicht bis zum 8ten allgemeinen Concil incl. Er wurde im 16ten Jahrhundert von Andreas Darmarius aus Morea gebracht, von Pappus, einem Strasburger Theologen, gekauft, und von ihm im I. 1601 mit lateinischer Uebersetzung zuerst edirt. Später ging er auch in die Conciliensammlungen ueber; namentlich liess ihn Harduin im 5ten Bande seiner Collect. Concil. p. 1491 abdrücken, während Mansi ihn in seine einzelnen Theile zerlegte, und jeden derselben an der zutreffenden Stelle (bei jeder einzelnen Synode) mittheilte."

of Eutyches, assembled a robber council, ejected and banished Flavian, and put Severus in his stead. He, called the Independent,[1] set out with two hundred apostate monks from Eleutheropolis for Constantinople, muttering threats against Macedonius. Now this man without conscience had sworn to Anastasius never to move against the holy Council of Chalcedon: he broke the oath, and anathematised it with an infamous council. So the emperor Anastasius had involved Macedonius of Constantinople in many accusations and expelled him from his see, and banished him to Gangræ. Not long after, having sent away both him and his predecessor Euphemius, under pretence that the patriarchs had arranged with each other to take refuge with the Goths, he slew them with the sword. But the heretic Timotheus, surnamed Kolon and Litroboulos,[2] he gave to the Church as being of one mind with himself and obedient to his counsels. This man called a most impious synod, and lifted up his heel against the holy Council of Chalcedon. In agreement with Severus, they sent their synodical letters together to Jerusalem. These not being received kindled Anastasius to anger. So he banished Elias from the holy city to Evila and put John in his see, and sent thither the synodical acts of Severus and Timotheus."

The emperor Anastasius, whose dealings with the eastern patriarchs in his empire are thus described, reigned for 27 years, from 491 to 518. It is to him that, in the long contest which we are following, the

---

[1] $ἀκέφαλος$.   [2] Words of infamous meaning.

four Popes, Gelasius, Anastasius, Symmachus, and Hormisdas, have to direct their letters, their exhortations, and their admonitions. During the whole of this time, from 493, when the conflict between Odoacer and Theodorick is terminated, they will have exchanged the local rule of the Arian Herule for that of the Arian Ostrogoth. All write under what a pope of our own day has called " hostile domination". They write from the Lateran Patriarcheium, not, as St. Leo I., under the guardianship of one branch of the Theodosian house at Rome to another branch at Constantinople, but to eastern emperors, the first of their line who openly assume the right to dictate to Catholics what they are to believe. Zeno, Basiliscus, and Anastasius found patriarchs, who could sanction by their subscription much greater violations of all Christian right than St. Athanasius had denounced in Constantius, and St. Basil in Valens. They found, also, five Popes in succession, living themselves "under hostile domination," who resisted their tyranny, and saved both the doctrine and the discipline of the Church. Without these Popes it is plain that the Council of Chalcedon would have been given up in the East, and the Eutychean heresy made the doctrine of the eastern Church.

We have seen the courageous act of the patriarch Euphemius in refusing absolutely to crown Anastasius, whom he suspected to be an Eutychean, until he had received a written declaration from him that he would maintain the Council of Chalcedon. In the first three years of his reign, Anastasius gained popularity by enacting wise laws, and by removing a severe and detested tax, so that, in the words of the ancient biographer of St. Theodore, "what was to become a field of destruction appeared a paradise of pleasure".[1]

As soon as Gelasius became Pope, Euphemius sent him, according to custom, synodal letters. He assured the Pope of his true faith. He recognised in him the divinely appointed head of the Church. We have the answer of the Pope to his letter, and as this recognition on the part of the bishop immediately following Acacius is all-important, it will be well to quote the very words which show it.[2] "You have read," writes Pope Gelasius to Euphemius, "the sentence, 'Faith comes by hearing, and hearing by the word of God'; that word, for instance, by which He promised that the gates of hell should never prevail over the confession of the blessed Apostle Peter. And, therefore, you thought, with reason, because God is faithful in His words, unless He had promised to institute some such thing, He would not bring about a true fulfilment of His promise. Then you say that we, by the grace of the Divine Providence, as He (*i.e.*, Christ) pointed out, do not fail in charity to the holy churches because Christ has placed me in the pontifical seat, not needing, as he says, to be taught, but understanding all things necessary for the unity of the Church's body. I, indeed, personally, am the least of all men, most unworthy for the office of such a see, except that supernal grace ever works great things out of small. For what should I think of

[1] Civiltà, vol. iii., 1855, p. 429. Acta SS. Jan. XI. [2] Mansi, viii. 5. *Ep.* i.

myself, when the Teacher of the nations declares himself the last, and not worthy to be called an apostle. But to return to your words; if you have with truth ascertained that these gifts have been conferred on me by God, which, whatever goods they are, are gifts of God, follow then the exhortation of one who needs not to be taught, of one who, by supernal disposition, keeps watch over all things which touch the unity of the churches, and, as you assert, offers a bold resistance to the devil, the disturber of true peace and the structure which contains it. If, then, you pronounce that I am in possession of such privileges, you must either follow what you assert to be Christ's appointment, or, which God forbid, show yourself openly to resist the ordinances of Christ, or you throw out such things about me for the pleasure of making a show."[1]

Euphemius[2] complained that the election of the new Pope had not been communicated to him, as was usual. He besought indulgence in respect of the conditions imposed on him, since the people of Constantinople would not endure the expulsion of Acacius from the diptychs. The Pope should rather forgive the dead, and himself write to the people. To this the Pope replied: "Truly that was an old Church rule with our fathers, by whom the one Catholic and apostolic communion was preserved free from every pollution by those who desired it. But now, when you prefer strange companionship before the return to a pure and

[1] Ad veniam luxuriæ de me cognosceris ista jactare.
[2] See Photius, i. 129-130. Civiltà Cattolica, vol. iii., 1855, pp. 524-5.

blameless union with St. Peter, how should we sing the Lord's song in a strange land? How should we offer the old bond of the apostolic ordinance to men who belong to another communion, and prefer to it, according to your own testimony, condemned heretics." Euphemius, then, is inconsistent: he must either admit to his own communion all who are in communion with heretics, or remove all. The excuse of necessity and fear of the people will not stand, and is unworthy of a bishop, who has to lead his people, not to be led by them; who has to account to God for his flock, while his flock have not to account for him. If Euphemius is afraid of men, the Pope is more afraid, but it is of the judgment of God.

But while, immediately after the death of Acacius, his successors, Fravita and Euphemius, were renouncing his pretensions, at the same time that they would not surrender his person, it is well to see how the bishops of eastern Illyricum, subjects of the emperor Anastasius, addressed the Pope upon his accession.

"Holy apostolic Lord and most blessed Father of fathers, we have received with becoming reverence the wholesome precepts of your apostolate, and return the greatest thanks to Almighty God and your Blessedness that you have deigned to visit us with pastoral admonition and evangelic teaching. For it is our desire and prayer to obey your injunctions in all things, and, as we have received from our fathers, to maintain without stain the precepts of the Apostolic See, which your life and merits have inherited, and to keep the orthodox religion, which you preach, with faithful and blameless devotion, so far as our rude perception allows. For, even before your injunction, we had avoided the communion of Peter, Acacius, and all his followers, as pestilent contagion; and much more now, after the admonition of the Holy See, must we abstain from that pollution. And if there be any others, who have followed, or shall follow, the sect of Eutyches or Peter and Acacius, or have anything to do with their accomplices and associates, they are to be entirely avoided by us, who seek a blameless obedience to the Apostolic See according to the divine commands and the statutes of the fathers. And if there be any, which we neither suppose nor desire, who, with bad intention, think it their duty to separate from the Apostolic See, we abjure their company, for, as we said, guarding in all things the precepts of the fathers, and following the inviolable rules of the holy canons, we strive with a common faith and devotion to obey that of your apostolic and singular see . . . and we beg your apostolate to send us some one from your angelical see, that in his presence arrangements may be made, according to the orthodox faith, and the fulfilling of your command."[1]

Several letters of Gelasius show that the privileges claimed by the Byzantine archbishop came frequently into discussion in the contest respecting the retention of the name of Acacius in the diptychs. Thus he finds it monstrous that they allege canons against which they are shown to have always acted by their illicit ambition.

[1] Mansi, viii. 13. Rescriptum episcoporum Dardaniæ ad Gelasium Papam.

"They[1] object canons to us, not knowing what they say, for these they break by the very fact that they decline to obey the first see when it gives sound and good advice. It is the canons themselves which order appeals of the whole Church to be brought to the examination of this see. But they have never sanctioned appeal from it. Thus it is to judge of the whole Church, but itself to go before no judgment. Never have they enjoined judgment to be passed on its judgment; but have made its sentence indissoluble, as its decrees are to be followed. . . . Should the bishop of Constantinople, who according to the canons holds no rank among bishops, not be deposed when he falls into communion with false believers?" No place among bishops, because the canon of 381 and the canons of 451 had not been received. Thus, in his great letter[2] to all the Illyrian bishops, he asks: "Of what see was he bishop? Of what metropolitan church was he the prelate? Was it not of a church the suffragan of Heraclea? We laugh at the claim of a prerogative for Acacius because he was bishop of the imperial city. Did not the emperor often hold his court at Ravenna, at Milan, at Sirmium, at Treves? Did the bishops of these cities ever claim to themselves a dignity beyond the measure of that which had descended to them from ancient times? Can Acacius show that he acted by any council in excluding from Alexandria John, a Catholic consecrated by Catholics; in putting in Peter, a detected and condemned heretic,

---

[1] *Ep.* iv. *ad Faustum;* Mansi, viii. 17.
[2] *Ep.* xiii. *Valde mirati sumus;* Mansi, viii. 49.

without consulting the Apostolic See? In boldly assuming the power to expel Calendion from Antioch, and, without knowledge of the Apostolic See, put in again the heretic Peter, who had been condemned by himself? Certainly if the rank of cities is considered, that of the bishops of the second and third see is greater than that of the see which not only holds no rank among bishops, but has not even the rights of a metropolitan. The power of the secular kingdom is one thing, the distribution of ecclesiastical dignities is another. The smallness of a city does not diminish the rank of a king residing in it; nor does the imperial presence change the measure of religious rank. Let that city be renowned for the power of the actual empire; but the strength, the liberty, the advance of religion under it consists in religion holding its own undisturbed measure in the presence of that power." Then he refers to the fact how, forty years before, the emperor Marcian himself interceded with Pope Leo to increase the dignity of that see, but could obtain nothing against the rules; and then gave the highest praise to St. Leo, because nothing would induce him to violate the canons, and to the other fact that Anatolius, himself bishop of Constantinople, confessed that it was rather his clergy than himself who made this attempt, and that all lay in the power of the Apostolic See. And, thirdly, did not St. Leo, who confirmed the Council of Chalcedon, annul in it whatever was done beyond the Nicene canons? If it was said that, in the case of the bishops of Alexandria and of Antioch, it was rather the emperor who had acted than Acacius, should not a bishop suggest to a Christian prince, whose favour he enjoyed to the utmost, that he should suffer the Church to keep her own rules, and judgment on bishops should be given by bishops in council. If a bishop was the greater for being bishop of the imperial city, should he not be the more courageous in suggesting the right course? Then he quotes Nathan before David, and St. Ambrose before Theodosius, and St. Leo reproving the second Theodosius for excess of power in the case of the Latrocinium of Ephesus; and Pope Hilarus reproving the emperor Anthemius, and Pope Simplicius and Pope Felix resisting not only the tyrant Basiliscus, but the emperor Zeno, and they would have succeeded if he had not been urged on by the bishop of Constantinople. "And we also," adds the Pope, "when Odoacer, the barbarian and heretic, held the kingdom of Italy, when he commanded us to do wrong things, by the help of God, as is well known, did not obey him."

In this same letter the Pope uses the following words: "We are confident that no one truly a Christian is ignorant that the first see, above all others, is bound to execute the decree of every council which the assent of the universal Church has approved; for it confirms every council by its authority, and maintains it by its continued rule, in virtue of its own principate which the blessed Apostle Peter received by the voice of the Lord, but continues to hold and retain by the Church subsequently following it".

Pope Gelasius had in vain striven to gain the emperor Anastasius. After the return of his legates,

Faustus and Irenæus, who had gone in the embassy of Theodorick to Constantinople, he wrote to the emperor, in the year 494, a famous letter,[1] warning him to defend the Catholic faith, which Anastasius had not yet openly deserted, nor professed himself an Eutychean. In it he says: "Glorious son, as a Roman born, I love, I reverence, I receive you as Roman emperor: as holder, however unworthy, of the Apostolic See, I endeavour as best I can to supply by opportune suggestions whatever I find wanting to the complete Catholic faith. For a dispensation of the divine word has been laid upon me; woe is me if I preach not the Gospel! Since the blessed Apostle Paul, the vessel of election, in his fear thus cries out, how much more have I in my smallness to fear if I shrink from the ministry of preaching inspired by God, and transmitted to me by the devotion of the fathers? I entreat your piety not to take for arrogance the execution of a divine duty.[2] Let not a Roman prince esteem the intimation of truth in its proper sense an injury. Two, then, O emperor, there are by whom this world is ruled in chief—the sacred authority of pontiffs and the royal power. Of these that of priests weighs the heavier, insomuch as they will have in the divine judgment to render an account for kings themselves. For you know, most gracious son, that pre-eminent as you are in dignity over the human race, you nevertheless bow the neck submissively to those who preside over things divine. From them you seek the terms of salvation;

---

[1] Mansi, viii. 30-5.   [2] Ne arrogantiam judices divinæ rationis officium.

and you recognise that it is your duty in the order of religion to submit rather than to command in what concerns the reception and the distribution of heavenly sacraments. As to these matters, then, you know that you depend on their judgment, and do not wish them to be controlled by your will. For if, in what regards the order of public discipline, the ministers of religion, recognising that empire has been conferred on you by a disposition from above, obey your laws, lest they should appear to oppose a sentence issued merely in worldly matters, with what affection ought you to obey those who are appointed for the distribution of venerable mysteries? Moreover, as no slight responsibility lies upon pontiffs, if in the worship of God they are silent as to what is fitting, so for rulers it is no slight danger if, when bound to obey, they show contempt. And if the hearts of the faithful should submit as a general rule to all bishops when rightly treating divine things, how much more is consent to be given to the prelate of that see whom the will of God Himself has made pre-eminent over all bishops, and the piety of the whole Church continuously following it out has acknowledged?[1] Herein you evidently perceive that no one by mere human counsel can ever raise himself to the privilege or confession of him whom the voice of Christ set over all, whom the Church we venerate has always confessed and devotedly holds to be her Primate. Human presumption may attack the appointments of

---

[1] Quem cunctis sacerdotibus et Divinitas summa voluit præeminere, et subsequens Ecclesiæ generalis jugiter pietas celebravit.

divine judgment; but no power can succeed in overthrowing them. Do not, I entreat, be angry with me if I love you so well as to wish you to possess for ever the kingdom which has been given to you in time, and that, having empire in the world, you should reign with Christ. You do not allow anything to perish in your own laws, nor loss to be inflicted on the Roman name. With what face will you ask of Him rewards *there* whose losses *here* you do not prevent? One is my dove, my perfect is one; one is the Christian, which is the Catholic faith. There is no cause why one should allow any contagion to creep in; for 'he who offends in one is guilty of all,' and 'he who despises small things perishes by little and little'. This is that against which the Apostolic See provides with the utmost care. For since the Apostle's glorious confession is the root of the world, it must not be touched by any rift of pravity, nor suffer the least spot. For if—may God avert a thing which we are sure is impossible—any such thing were to happen, how could we resist any error?—how could we correct those who err? If you declare that the people of one city cannot be composed to peace, what should we make of the whole world's universe were it deceived by our prevarication? The series of canons coming down from our fathers, and a multifold tradition, establish that the authority of the Apostolic See is set for all Christian ages over the whole Church. O emperor, if anyone made any attempt against the public laws, you could not endure it; do you think it is of no concern to your conscience that the people subject to you may purely and sincerely worship God? Lastly, if it is thought that the feeling of the people of one city should not be offended by the due correction of divine things, how much more neither may we, nor can we, by offence of divine things injure the faith of all who bear the Catholic name?"

How distinctly, and with what unfaltering conviction, the Pope of 494, then locally a subject of Theodorick the Arian, set forth to the emperor at Constantinople the universal authority of the Holy See, grounded on what he calls the Apostle's glorious confession, on which followed the Divine Word creating his office, is apparent through the whole of this magnificent letter. Moreover, the distinction of the Two Powers and the character of their relation to each other, and the divine character of each as a delegation from God, solemnly uttered by the Pope Gelasius in 494 to the Roman emperor so unworthy of the rank which the Pope recognised in him, have passed into the law and practice of the Church during the 1400 years which have since run out, and will form part of it for ever. Anastasius disregarded all that the Pope said. He persecuted to the utmost his bishop Euphemius, because, though not admitted to communion by the Pope, inasmuch as he refused to erase from the diptychs the name of Acacius, he yet vigorously maintained the decrees of the Council of Chalcedon. At length the emperor, having ended his Isaurian wars and sufficiently strengthened the Monophysite party, succeeded in deposing him in 496.

His instruments in this were the cowardly court bishops,[1] ready to be moved to anything, who had also on this occasion to confirm the Henotikon of Zeno. Euphemius was banished to Paphlagonia. The people rioted in the circus and demanded his restoration, but in vain. However, they always venerated him as a saint. While the emperor Anastasius was deposing at Constantinople the bishop who withstood and reproved his conduct in supporting the Eutychean heresy, while also he was compelling the resident council not only to depose the bishop, but to confirm the document, originally drawn up by Acacius, forced upon the bishops of his empire by Zeno, and now again forced upon them by Anastasius, Gelasius was holding a council of seventy bishops at Rome. What he enacted there synodically is a proof of the entirely different spirit which prevailed in the independent West. Here Pope and bishops alike were living under hostile domination, that of Arian governments, but they were not crouching before the throne of a despot. The Pope and the bishops passed at the synod of 496 the following decrees:

"After the writings of the Prophets, the gospels, and the Apostles, on which by the grace of God the Catholic Church is founded, this also we have judged fit to be expressed: Although all the Catholic churches spread throughout the world are the one bridal-chamber of Christ, nevertheless the holy Roman Church has been set over all other churches, by no constitution of a

[1] Photius, 134; Hefele, *C. G.*, ii. 597.

council, but obtained the Primacy by the voice of our Lord in the Gospel : 'Thou art Peter,' &c.

"To whom was also given the companionship of the most blessed Apostle Paul, the vessel of election, who, not at another time, as heretics battle, but on one and the same day with Peter combating in the city of Rome under the emperor Nero, was crowned. And they consecrated this holy Roman Church to Christ the Lord, and by their presence and worshipful triumph set it over all the churches in the world.

"First, therefore, is the Roman Church, the see of the Apostle Peter, having neither spot, nor wrinkle, nor any such thing.

"Second is the see consecrated at Alexandria in the name of blessed Peter by Mark, his disciple, the Evangelist. And he, sent by the Apostle Peter to Egypt, preached the word of truth, and consummated a glorious martyrdom.

"Third is the see of the same most blessed Apostle Peter held in honour at Antioch, because there he dwelt before he came to Rome, and there first the name of Christian was given to the new people.

"And though no other foundation can be laid, save that which is laid, Jesus Christ, yet the said Roman Church, after those writings of the Old or New Testament, which we receive according to rule, does also not prohibit the following : that is, the holy Nicene Council, of three hundred and eighteen fathers, held under the emperor Constantine; the holy Council of Ephesus, in which Nestorius was condemned, with the consent of Pope Cœlestine, under Cyril, bishop of Alexandria, and Arcadius, sent from Italy; the holy Council of Chalcedon, held under the emperor Marcian and Anatolius, bishop of Constantinople, in which the Nestorian and Eutychean heresies were condemned, with Dioscorus and his accomplices."[1]

Thus, twelve years after the attempt of Acacius to set himself up independent of Rome, and while his next two successors were soliciting the recognition of Rome, but at the same time were refusing to surrender his person to condemnation, a Council at Rome pulled down the whole scaffolding on which the pretension of Acacius had been built.

For while this council omitted from the list of councils acknowledged to be general that held at Constantinople in 381, it likewise proclaimed the falsity of the ground alleged in the canon passed in that council, which gave to Constantinople the second rank in the episcopate because it was New Rome, which canon again was enlarged by the attempt at the Council of Chalcedon to put upon the world the positive falsehood asserted in the rejected 28th canon, that the fathers had given its privileges to the Roman See because it was the imperial city.

The significance of this decree at such a time cannot be exaggerated. While the emperor's own Church and

---

[1] Hefele, *C. G.*, ii. 597-605, has most carefully considered the text and the date of the Council of 496. I have followed him in his choice of the text of the best manuscripts, and inasmuch as the biblican canon—the same as that held in the African Church about 393—seems to have been confirmed by Pope Hormisdas somewhat later, I have not made use of it in this place.

bishop are separated by a schism from the Pope, while the Pope recognises the emperor as the sole "Roman prince," and in that capacity speaks of him as "preeminent in dignity over the human race," he states at the head of a council, in the most peremptory terms, that the Principate of Rome is of divine institution, *not* the constitution of any council. The decree thus passed is a formal contradiction of the 28th canon which St. Leo had, forty years before, rejected.

When we come to the termination of the schism this fact is to be borne in mind as being accepted voluntarily by those whom it specially concerned, and whose actions during a hundred years immediately preceding it condemned. For the decree, besides, does not acknowledge the see of Constantinople as patriarchal. Acacius had been appointing those who were really patriarchs: here his own pretended patriarchate is shown to be an infringement on the ancient order of the Church. Here the Pope in synod, as before in his letter to the Illyrian bishops, declares of the see of Constantinople that "it holds no rank among bishops".

And, again, the Roman Council, in all its wording, censures the bishops who had been so weak as to accept a decree upon the faith of the Church from the hand of emperors, first the usurper Basiliscus, then Zeno, and at the time itself Anastasius. And under this censure lay not only Acacius, but the three following bishops of Constantinople—Fravita, Euphemius, and Macedonius. For though the last two were firm enough to suffer deposition, and afterwards death, for the faith of Chalcedon, they were not firm enough to refuse the emperor's imposition of an imperial standard in doctrine, the acceptance of which would have destroyed the essential liberty of the Church.

Two months after the violent deposition of Euphemius at Constantinople, Pope Gelasius closed a pontificate of less than five years, in which he resisted the wickedness and tyranny of Anastasius, as Pope Felix had resisted the like in Zeno. Space has allowed me to quote but a few passages of the noble letters which he has left to the treasury of the Church. It may be noted that with his pontificate closes the period of about twenty years, from 476 to 496, in which no single ruler of East or West, great or small, professed the Catholic faith. The eastern emperors were Eutychean; the new western rulers Arian, save when they were pagan. The next year the conversion of Clovis, with his Franks, opens a new series of events. We may allow Gelasius,[1] in his letter to Rusticus, bishop of Lyons, to express the character of his time. "Your charity, most loving brother, has brought us great consolation in the midst of that whirlwind of calamities and temptations under which we are almost sunk. We will not weary you by writing how straitened we have been. Our brother Epiphanius (bishop of Ticinum or Pavia) will inform you how great is the persecution we bear on account of the most impious Acacius. But we do not faint. Under such pressure neither courage fails nor zeal.

[1] *Epist.* xviii.

Distressed and straitened as we are, we trust in Him who with the trial will find an issue, and if He allows us for a time to be oppressed, will not allow us to be overwhelmed. Dearest brother, see that your affection, and that of yours, to us, or rather to the Apostolic See, fail not, for they who are fixed into the Rock with the Rock shall be exalted."[1]

[1] Qui enim in petra solidabuntur cum petra exaltabuntur.

CHAPTER III.

PETER STOOD UP.

SEVEN days after the death of Gelasius, Anastasius, a Roman, ascended the apostolic throne, which he held from November, 496, to November, 498. We have two letters from him extant, both important. In that addressed upon his own accession, which he sent to the emperor Anastasius by the hands of Germanus, bishop of Capua, and Cresconius, bishop of Trent, on occasion of Theodorick's embassy for the purpose of obtaining the title of king, he strove to preserve the "Roman prince" from the Eutychean heresy.

"I announce to you the beginning of my pontificate, and consider it a token of the divine favour that I bear the same as your own august name. This is an assurance that, like as your own name is pre-eminent among all the nations in the world, so by my humble ministry the See of St. Peter, as always, may hold the Principate assigned to it by the Lord God in the whole Church. We therefore discharge a delegated office in the name of Christ."[1] After beseeching the emperor that the name of Acacius should be effaced, in which he is carrying out the judgment of his predecessor,

[1] *Epist.* i.; Labbé, v. 406.

Pope Felix, he mentions the full instructions given to his legates, in order that the emperor might plainly see how, in that matter, the sentence of the Apostolic See had not proceeded from pride, but rather had been extorted by zeal for God as the result of certain crimes. "This we declare to you, in virtue of our apostolic office, through special love for your empire, that, as is fitting, and the Holy Spirit orders, obedience be yielded to our warning, that every blessing may follow your government. Let not your piety despise my frequent suggestion, having before your eyes the words of our Lord, 'He who hears you, hears Me: and he who despises you, despises Me: and he who despises Me, despises Him who sent Me'. In which the Apostle agrees with our Saviour, saying, 'He who despises these things, despises not man but God, who has given us His Holy Spirit'. Your breast is the sanctuary of public happiness, that through your excellency, whom God has ordered to rule on earth as His Vicar, not the resistance of hard pride be offered to the evangelic and apostolic commands, but an obedience which carries safety with it."

The Pope, then, standing alone in the world, and locally the subject of Theodorick the Goth, makes the position of the Roman emperor in the world, and the Pope in the Church, parallel to each other. Both are divine legations. The Pope, speaking on divine things, claims obedience as uttering the will of the Holy Spirit, which Pope Anastasius asserts, just as Pope Clement I., five hundred years before, had asserted it, in the first pastoral letter which we possess. He, living on sufferance in Rome, asserts it to the despotic ruler of an immense empire, throned at Constantinople, in reference to a bishop of Constantinople, whose name he requires the emperor to erase from the sacred records of the Church as a condition of communion with the Apostolic See.

This letter was directed to the East, the other belongs to the West, and records an event which was to affect the whole temporal order of things in that vast mass of territories already occupied by the northern tribes. On Christmas day of the year 496, that is, one month after the accession of Pope Anastasius, the haughty Sicambrian bent his head to receive the holy oil from St. Remigius, to worship that which he had burnt, and to burn that which he had worshipped. Clovis, chief of the Franks, and a number of his warriors with him, were baptised in the name of the most holy Trinity, never having been subject to the Arian heresy. Upon that event, the Holy See no longer stood alone, and the ring of Arian heresy surrounding it was broken for ever. The words of the Pope are these:

"Glorious son, we rejoice that your beginning in the Christian faith coincides with ours in the pontificate. For the See of Peter, on such an occasion, cannot but rejoice when it beholds the fulness of the nations come together to it with rapid pace, and time after time the net be filled, which the same Fisherman of men and blessed Doorkeeper of the heavenly Jerusalem was bidden to cast into the deep. This we have wished to signify to your serenity by the priest Eumerius, that, when you hear of the joy of the father in your good works, you may fulfil our rejoicing, and be our crown, and mother Church may exult at the proficiency of so great a king, whom she has just borne to God. Therefore, O glorious and illustrious son, rejoice your mother, and be to her as a pillar of iron. For the charity of many waxes cold, and by the craftiness of evil men our bark is tossed in furious waves, and lashed by their foaming waters. But we hope in hope against hope, and praise the Lord, who has delivered thee from the power of darkness, and made provision for the Church in so great a prince, who may be her defender, and put on the helmet of salvation against all the efforts of the infected. Go on, therefore, beloved and glorious son, that Almighty God may follow with heavenly protection your serenity and your realm, and command His angels to guard you in all your ways and to give you victory over your enemies round about you."[1]

Towards the end of the sixth century, the Gallic bishop, St. Gregory of Tours, notes how wonderfully prosperity followed the kingdom which became Catholic, and contrasts it with the rapid decline and perishing away of the Arian kingdoms. And, indeed, this letter of the Pope may be termed a divine charter, commemorating the birthday of the great nation, which led the way, through all the nations of the West, for

[1] Mansi, viii. 193.

their restoration to the Catholic faith, and the expulsion of the Arian poison. No one has recorded, and no one knows, the details of that conversion, by which the Church, in the course of the sixth century, recovered the terrible disasters which she had suffered in the fifth; a conversion by which the sturdy sons of the North, from heretics, became faithful children, and by which she added the Teuton race, in all its new-born vigour and devotion, to those sons of the South, whose conversion Constantine crowned with his own. St. Gregory of Tours calls Clovis the new Constantine, and in very deed his conversion was the herald of a second triumph to the Church of God, which equals, some may think surpasses even, the grandeur of the first.

It was fitting that the See of Peter should sound the note, which was its prelude, by the mouth of Anastasius, as the pastoral staff of St. Gregory was extended over its conclusion.

Scarcely less remarkable than the words of Pope Anastasius were those addressed to the new convert by a bishop, the temporal subject of the Burgundian prince, Gundobald, an Arian, that is, by St. Avitus of Vienna, grandson of the emperor of that name. Before the baptismal waters were dry on the forehead of the Frankish king, he wrote to him in these words:[1]

"The followers of all sorts of schisms, different in their opinions, various in their multitude, sought, by

[1] Epistola Aviti episcopi Viennensis ad Clodoveum regem Francorum.—Mansi, viii. 175.

pretending to the Christian name, to blunt the keenness of your choice. But, while we entrust our several conditions to eternity, and reserve for the future examination what each conceives to be right in his own case, a bright flash of the truth has descended on the present. For a divine provision has supplied a judge for our own time. In making choice for yourself, you have given a decision for all. Your faith is our victory. In this case most men, in their search for the true religion, when they consult priests, or are moved by the suggestion of companions, are wont to allege the custom of their family, and the rite which has descended to them from their fathers. Thus making a show of modesty, which is injurious to salvation, they keep a useless reverence for parents in maintaining unbelief, but confess themselves ignorant what to choose. Away with the excuse of such hurtful modesty, after the miracle of such a deed as yours. Content only with the nobility of your ancient race, you have resolved that all which could crown with glory such a rank should spring from your personal merit. If they did great things, you willed to do greater. Your answer to that nobility of your ancestors was to show your temporal kingdom; you set before your posterity a kingdom in heaven. Let Greece exult in having a prince of our law; not that it any longer deserves to enjoy alone so great a gift, since the rest of the world has its own lustre. For now in the western parts shines in a new king a sunbeam which is not new. The birthday of our Redeemer fitly marked its bright rising. You were regenerated to salvation from the water on the same day on which the world received for its redemption the birth of the Lord of heaven. Let the Lord's birthday be yours also: you were born to Christ when Christ was born to the world. Then you consecrated your soul to God, your life to those around you, your fame to those coming after you.

"What shall I say of that most glorious solemnity of your regeneration? I was not able to be present in body: I did not fail to share in your joy. For the divine goodness added to these regions the pleasure that the message of your sublime humility reached us before your baptism. Thus that sacred night found us in security about you. Together we contemplated that scene, when the assembled prelates, in the eagerness of their holy service, steeped the royal limbs in the waters of life; when the head, before which nations tremble, bowed itself to the servants of God; when the helmet of sacred unction clothed the flowing locks which had grown under the helmet of war; when, putting aside the breastplate for a time, spotless limbs shone in the white robe. O most highly favoured of kings, that consecrated robe will add strength hereafter to your arms, and sanctity will confirm what good fortune has hitherto bestowed. Did I think that anything could escape your knowledge or observation, I would add to my praises a word of exhortation. Can I preach to one now complete in faith, that faith which he recognised before his completion? Or humility to one who has long shown us devotion, which now his profession claims as a debt? Or mercy to one whom a captive people, just set free by you, proclaims by its rejoicing to the world, and by its tears to God. In one thing I should wish an advance. This is, since through you God will make your nation all His own, that you would, from the good treasure of your heart, provide the seeds of faith to the nations beyond you, lying still in their natural ignorance, uncorrupted by the germs of false doctrine. Have no shame, no reluctance, to take the side of God, who has so exalted your side, even by embassies directed to that purpose. . . . You are, as it were, the common sun, in whose rays all delight; the nearest the most, but somewhat also those further off. . . . Your happiness touches us also; when you fight, we conquer."

It is easy to look back on the course of a thousand years, and see how marvellously these words, uttered by St. Avitus at the moment Clovis was baptised, were fulfilled in his people. "Your happiness touches us also; when you fight, we conquer." So spoke a Catholic bishop at the side, and from the court, of an Arian king, and thus he expressed the work of the Catholic bishops throughout Gaul in the sixth century then beginning. An apostate from the Catholic faith has said of them that they built up France as bees build a hive; but he omitted to say that they were able and willing to do this because they had a queen-bee at Rome, who, scattered as they were in various transitory kingdoms under heretical sovereigns, gave unity to all their efforts, and planted in their hearts the assurance of one undying kingdom. We shall have presently to quote other words of St. Avitus, speaking, as he says, in the name of all his brethren to the senators of Rome: "If the Pope of the city is called into question, not one bishop, but the episcopate, will seem to be shaken". But that, which he here foresaw, explains in truth a process, of which we do not possess a detailed history, but which resulted, by the time of St. Gregory, in the triumph of the Catholic faith over that most fearful heresy which had contaminated the whole Teuton race of conquerors at the time of their conquest. The glory of this triumph is divided between St. Peter's See and the Catholic bishops in the several countries, working each in union with it. So was formed the hive, not only of France, but of Christ; the hive which nurtured all the nations of the future Europe.

When Faustus,[1] the ambassador sent by Theodorick to Anastasius to obtain for him the royal title, returned to Rome in 498, he found Pope Anastasius dead. The deacon Symmachus was chosen for his successor, and his pontificate lasted more than fifteen years. But Faustus had hoped to gain the approval of Pope Anastasius to the Henotikon set up by the emperor Zeno at the instance of Acacius, and forced by the emperor Anastasius on his eastern bishops, and specially on three successive bishops of Constantinople—

---

[1] See for this narrative the German Röhrbacher, viii. 486; Civiltà, 1855, art. 9, pp. 152-3; Hefele, ii. 607; Photius, i. 136.

Fravita, Euphemius, and Macedonius—who took the place of the second, when he had been expelled by the emperor. Faustus, who was chief of the senate, with a view to gain to the emperor's side the Pope to be elected in succession to Anastasius, brought from the East the old Byzantine hand; that is to say, he bore gifts for those who could be corrupted, threats for those who could be frightened, and deceit for all. So freighted he managed to bring about a schism in the papal election, and the candidate whom he favoured, Laurentius, was set up by a smaller but powerful party against the election of Symmachus. Thus disunion was introduced among the Roman clergy, which brought about, during the five succeeding years, many councils at Rome, and embarrassed the action of the Pope more than the Arian government of Theodorick.[1] The difficulty of the times was such that, instead of holding a synod of bishops at Rome to determine which election was valid, the two candidates, Symmachus and Laurentius, went to Ravenna, and submitted that point to the decision of the king Theodorick, Arian as he was. That decision was that he who was first ordained, or who had the majority for him, should be recognised as Pope; Symmachus fulfilled both conditions, and his election was acknowledged.

Symmachus, in the first year of his pontificate, 499,

---

[1] Photius, i. 137. Der Einfluss des römischen Stuhles war doch mehr durch die Erneuerung des laurentianischen Schisma als durch die Macht der arianischen Ostgothen auf längere Zeit gelähmt.

addressed to the Roman emperor, in his Grecian capital, a renowned letter, termed "his defence" against imperial calumnies. This letter alone would be sufficient to exhibit the whole position of the Pope in regard to the eastern emperor at the close of the fifth century. Space allows me to quote only a part of it.

The emperor of Constantinople was very wroth at the frustration of his plan to get influence over the Pope by the appointment of Laurentius, and reproached Pope Symmachus with moving the Roman senate against him. The Pope replied:[1]

"If, O emperor, I had to speak before outside kings, ignorant altogether of God, in defence of the Catholic faith, I would, even with the threat of death before me, dwell upon its truth and its accord with reason. Woe to me if I did not preach the gospel. It is better to incur loss of the present life than to be punished with eternal damnation. But if you are the Roman emperor, you are bound kindly to receive the embassies of even barbarian peoples. If you are a Christian prince, you are bound to hear patiently the voice of the apostolic prelate, whatever his personal desert.[2] I must confess that I cannot pass over, either on your account or on my own, the point whether you issue with a religious mind against me the insults which you utter in presence of the divine judgment. Not on my own account, when I remember the Lord's promise, 'When they persecute you, and say all manner of evil against

---

[1] *Ep.* vi. ; Mansi, viii. 213-217.
[2] Qualiscunque præsulis apostolici debes vocem patienter audire.

you, for justice' sake, rejoice'. Not on your account, because I wish not a result to my own glory, which would weigh heavily upon you. And being trained in the doctrine of the Lord and the Apostles, I am anxious to meet your maledictions with blessing, your insults with honour, your hatred with charity. But I would beg you to reflect whether He who says, 'Vengeance is Mine, I will repay,' will not exact the more from you for my forbearance. . . . I wish, then, that the insults, which you think proper to bestow on my person, while they are glorious to me, may not press upon you. To my Lord it was said by some: 'Thou hast a devil; a man that is a glutton, born of fornication'. Am I to grieve over such things? Divine and human laws present the condition to him who utters them: 'In the mouth of two or three witnesses every word shall stand'. O emperor, what will you do in the divine judgment? Because you are emperor, do you think there is no judgment of God? I pass over that it becomes not an emperor to be an accuser. Again, both by divine and human laws, no one can be at once accuser and judge. Will you plead before another judge? Will you stand by him as accuser? You say I am a Manichean. Am I an Eutychean, or do I defend Eutycheans, whose madness is the chief support [1] to the Manichean error? Rome is my witness, and our records bear testimony, whether I have in any way

---

[1] *I.e.*, Manicheans placed the seat of evil in matter, and Eutycheans denied the materiality of the Lord's body. The Pope alludes to the Emperor's Eutychean doctrine.

deviated from the Catholic faith, which, coming out of paganism, I received in the See of the Apostle St. Peter. . . . Is it because I will offer no acceptance to Eutycheans? Such reproaches do not wound me, but they are a plain proof that you wished to prevent my advancement, which St. Peter by his intervention has imposed. Or, because you are emperor, do you struggle against the power of Peter? And you, who accept the Alexandrian Peter, do you strive to tread under foot St. Peter the Apostle in the person of his successor, whoever he may be? Should I be well elected if I favoured the Eutycheans? if I held communion with the party of Acacius? Your motive in putting forward such things is obvious. Now, let us compare the rank of the emperor with that of the pontiff. Between them the difference is as great as the charge of human and divine things. You, emperor, receive baptism from the pontiff, accept sacraments, request prayers, hope for blessing, beg for penitence. In a word, you administer things human, he dispenses to you things divine. If, then, I do not put his rank superior, it is at least equal. And do not think that in mundane pomp you are before him, for 'the weakness of God is stronger than men'. Consider, then, what becomes you. But when you assume the accuser's part, by divine and human law you stand on the same level with me; in which, if I lose the highest rank, as you desire, if I be convicted by your accusation, you will equally lose your rank if you fail to convict me. Let the world judge between us, in the sight of God and His angels; let us be a spectacle for every age, in which either the priest shall exhibit a good life, or the emperor a religious modesty. For the human race is ruled in chief by these two offices, so that in neither of them should there be anything to offend God, especially because each of these ranks would appear to be perpetual, and the human race has a common interest in both.

"Allow me, emperor, to say, Remember that you are a man in order to use a power granted you by God. For though these things pass first under the judgment of man, they must go on to the divine examination. You may say, It is written, 'Let every soul be subject to higher powers'. We accept human powers in their proper place until they set up their wills against God. But if all power be from God, more then that which is given to things divine. Acknowledge God in us and we will acknowledge God in thee. But if you do not acknowledge God, you cannot use a privilege derived from Him whose rights you despise. You say that conspiring with the senate I have excommunicated you. In that I have my part; but I am following fearlessly what my predecessors have done reasonably. You say the Roman senate has ill-treated you. If we treat you ill in persuading you to quit heretics, do you treat us well who would throw us into their communion? What, you say, is the conduct of Acacius to me? Nothing if you leave him. If you do not leave him it touches you. Let us both leave the dead. This is what we beg, that you have nothing to do with what Acacius did. Making your own what Acacius did, you accuse us of objections.

We avoid what Acacius did; do you avoid it also. Then we shall both be clear of him. Thus relinquishing his actions you may be joined with our cause, and be associated with our communion without Acacius. It has always been the custom of Catholic princes [1] to be the first to address the apostolic prelates upon their accession, and they have sought, as good sons, with the due affection of piety, that chief confession and faith to which you know that the care of the whole Church has been committed by the voice of the Saviour Himself. But since public circumstances may have caused you to omit this, I have not delayed to address you first, lest I should be thought to consider more my own private honour than solicitude for the whole flock of the Lord.

"You say that we have divulged your compelling by force those who had long kept themselves apart from the contagion of heresy to yield to its detestable communion. In this, O chief [2] of human powers, I, as successor, however unmerited, in the Apostolic See, cease not to remind you that whatever may be your material power in the world, you are but a man. Review all those who, from the beginning of the Christian belief, have attempted with various purpose to persecute or afflict the Catholic faith. See how those who used such violence have failed, and the orthodox truth

---

[1] Catholici principes quidem semper apostolicos præsules institutos suis literis prævenerunt, et illam confessionem fidemque præcipuam, tanquam boni filii, quæsierunt debitæ pietatis affectu, cui noscis ipsius Domini Salvatoris ore curam totius Ecclesiæ delegatam.

[2] Ubi te, rerum humanarum princeps, qualiscunque Sedis Apostolicæ vicarius contestari mea voce non desino.

prevailed through the very means by which it was thought to be overthrown. And as it grew under its oppressors, so it is found to have crushed them. I wonder if even human sense, especially in one who claims to be called Christian, fails to see that among these oppressors must be counted those who assault Christian confession and communion with various superstitions. What matters it whether it be a heathen or a so-called Christian who attempts to infringe the genuine tradition of the apostolic rule? Who is so blind that in countries where every heresy has free licence to exhibit its opinions he should deem the liberty of Catholic communion alone should be subverted by those who think themselves religious?"

"All Catholic princes," the Pope repeats, "either at their own accession, or on knowing the accession of a new prelate to the Apostolic See, immediately addressed their letters to it, to show that they were in union with it. Those who have not done so declare themselves aliens from it. Your own writings would justify us in so considering you if we did not from your assault and hostility avoid you, whether as enemy or judge . . . but the accomplice of error must persecute him who is its enemy."

Let this letter from beginning to end be considered as written by a Pope just after his election, the validity of which had been disputed by another candidate whom the emperor had favoured—by a Pope living actually under the unlimited power of an Arian sovereign, who was in possession of Italy, and who ruled in right of a conqueror, though he used his power generally with moderation and equity; further, that it was addressed to one who had become the sole Roman emperor, the overlord of the king, who had just besought of him the royal title; that it required him to cast aside his patronage of Eutychean heretics; to rescind from the public records of the Church the name of that bishop who had composed the document called the Henotikon, the very document which the emperor was compelling his eastern bishops to accept and promulgate as the confession of the Christian faith. And let the frankness with which the Pope appeals to the universally admitted authority of St. Peter's See be at the same time considered, with the official statement that the emperors were wont immediately to acknowledge the accession of a Pope[1] and attest their communion with him.

What was the answer which the eastern emperor made to this letter? He did not answer by denying anything which the Pope claimed as belonging to his see, but by rekindling the internal schism which had been laid to sleep by the recognition of Pope Symmachus. Before sending this letter, the Pope had held a council of seventy-two bishops in St. Peter's on March 1, 499, which made important regulations to prevent cabal and disturbance at papal elections such as had just taken place. This council had been subscribed by Laurentius himself,[2] and the Pope in compassion[3] had

---

[1] Ad eam sua protinus scripta miserunt ut *se docerent ejus esse consortes.*— Mansi, viii. 217.

[2] See Hefele, ii. 607 and 209.   [3] "Intuitu misericordiæ," says Anastasius.

given him the bishopric of Nocera. Now the emperor Anastasius, reproved for his misdeeds and misbelief by Pope Symmachus in the letter above quoted, caused his agents, the patrician Faustus and the senator Probinus, to bring grievous accusations against Symmachus and to set up once more Laurentius as anti-pope.[1] In their passionate enmity they did not scruple to bring their charge against Pope Symmachus before the heretical king Theodorick. The result of this attempt was that Rome, during several years at least, from 502 to 506, was filled with confusion and the most embittered party contentions. Theodorick was induced to send a bishop as visitor of the Roman Church, and again to summon a council of bishops from the various provinces of Italy to consider the charges brought against the Pope. During the year 501 four such councils were held in Rome, of which it may be sufficient to quote the last, the Synodus Palmaris.[2] Its acts say that they were by command of king Theodorick to pass judgment on certain charges made against Pope Symmachus. That the bishops of the Ligurian, Æmilian, and Venetian provinces, visiting the king at Ravenna on their way, told him that the Pope himself ought to summon the council, " knowing that in the first place the merit or principate of the Apostle Peter, and then the authority of venerable councils following out the commandment of the Lord, had delivered to his see a singular power in the churches, and no instance could be produced in

---

[1] Hefele, ii. 216.   [2] Mansi, viii. 247-252; Hefele, ii. 623-5.

which the bishop of that see in a similar case had been subjected to the judgment of his inferiors". To which king Theodorick replied that the Pope himself had by letter signified his wish to convene the council. Then the Synodus Palmaris, passing over a narration of what had taken place in the preceding councils, came to this conclusion: "Calling God to witness, we decree that Pope Symmachus, bishop of the Apostolic See, who has been charged with such and such offences, is, as regards all human judgment, clear and free (because for the reasons above alleged all has been left to the divine judgment); that in all the churches belonging to his see he should give the divine mysteries to the Christian people, inasmuch as we recognise that for the above-named causes he cannot be bound by the charges of those who attack him. Wherefore, in virtue of the royal command, which gives us this power, we restore all that belongs to ecclesiastical right within the sacred city of Rome, or without it, and reserving the whole cause to the judgment of God, we exhort all to receive from him the holy communion. If anyone, which we do not suppose, either does not accept this, or thinks that it can be reconsidered, he will render an account of his contempt to the divine judgment. Concerning his clergy, who, contrary to rule, left their bishop and made a schism, we decree that upon their making satisfaction to their bishop, they may be pardoned and be glad to be restored to their offices. But if any of the clergy, after this our order, presume to celebrate mass in any holy place in the Roman Church without leave of Pope Symmachus, let him be punished as schismatic."[1]

This was signed by seventy-six bishops, of whom Laurentius of Milan and Peter of Ravenna stood at the head; and the two metropolitans accompany their subscription with the words, "in which we have committed the whole cause to the judgment of God".[2]

When this document reached Gaul, the bishops there, being unable to hold a council through the division of the country under different princes, commissioned St. Avitus, bishop of Vienne, to write in his name and their own, and we have from him the following letter addressed to Faustus and Symmachus, senators of Rome :[3]

"It would have been desirable that we should, in person, visit the city which the whole world venerates, for the consideration of duties which affect us both as men and as Christians. But as the state of things has long made that impossible, we could wish at least to have had the security that your great body should learn from a report of the assembled bishops of Gaul the entreaties called forth by a common cause. But since the separation of our country into different governments deprives us also of that our desire, I must first entreat that your most illustrious Order may not take offence at what I write as coming from one person. For, urged not only by letters, but charges from all my Gallic brethren, I have undertaken to be the organ of

[1] *Acts of the Synodus Palmaris.*—Mansi, viii. 247-251.
[2] Hefele, ii. 624.   [3] Mansi, viii. 293-5. *Ep.* xxxi.   Migne, vol. lix, 248.

communicating to you what we all ask of you. Whilst we were all in a state of great anxiety and fear in the cause of the Roman Church, feeling that our own state was imperilled when our head was attacked, inasmuch as a single incrimination would have struck us all down without the odium which attaches to the oppression of a multitude, if it had overturned the condition of our chief, a copy of the episcopal decree was brought to us in our anxiety from Italy, which the bishops of Italy, assembled at Rome, had issued in the case of Pope Symmachus. This constitution is made respectable by the assent of a large and reverend council: yet our mind is, that the holy Pope Symmachus, if accused to the world, had a claim rather to the support than to the judgment of his brethren the bishops. For as our Ruler in heaven bids us be subject to earthly powers, foretelling that we shall stand before kings and princes in every accusation, so is it difficult to understand with what reason, or by what law, the superior is to be judged by his inferiors. The Apostle's command is well known, that an accusation against an elder should not be received. How, then, is it lawful to incriminate the Principate of the whole Church? The venerable council itself providing against this in its laudable constitution, has reserved to the divine judgment a cause which, I may be permitted to say, it had somewhat rashly taken up; mentioning, however, that the charges objected to the Pope had in no respect been proved, either to itself or to king Theodorick. In face of all which, I, myself a Roman senator, and a Christian bishop, adjure you (so may the God you worship grant prosperity to your times, and your own dignity maintain the honour of the Roman name to the universe in this collapsing world), that the state of the Church be not less in your eyes than that of the commonwealth; that the power which God has given to you may be also for our good; and that you have not less love in your Church for the See of Peter, than in your city for the crown of the world. If, in your wisdom, you consider the matter to its bottom, you will see that not only the cause carried on at Rome is concerned. In the case of other bishops, if there be any lapse, it may be restored; but if the Pope of Rome is endangered, not one bishop but the episcopate itself will seem to be shaken. You well know how we are steering the bark of faith amid storms of heresies, whose winds roar around us. If with us you fear such dangers, you must needs protect your pilot by sharing his labour. If the sailors turn against their captain, how will they escape? The shepherd of the Lord's sheepcot will give an account of his pastorship; it is not for the flock to alarm its own pastor, but for the judge. Restore, then, to us if it be not already restored, concord in our chief."

Even after this synod at Rome, the opponents of Symmachus did not cease their attempts. Clergy and senators sent in a new memorial to the king Theodorick, in favour of the anti-pope Laurentius, who returned to Rome in 502; and it was four years, during which several councils were held, before the schism was finally composed. Theodorick then commanded that all the churches in Rome should be given up to Pope Symmachus,[1] and he alone be recognised as its bishop.

Against the attacks made upon the fourth synod, which had dismissed the consideration of the charges against the Pope as beyond its competence, Ennodius, at that time a deacon, afterwards bishop of Pavia, wrote a long defence. This writing was read at the sixth synod at Rome, held in 503, approved, and inserted in the synodal acts. We may, therefore, quote one passage from it, as the doctrine which it was the result of all this schism to establish.[2] "God has willed the causes of other men to be terminated by men; He has reserved the bishop of that one see without question to His own judgment. It was His will that the successors of the Apostle St. Peter should owe their innocence to heaven alone, and show a spotless conscience to that most absolute scrutiny. Do not suppose that those souls whom God has reserved to His own examination have no fear of their judges. The guilty has with Him no one to suggest excuse, when the witness of the deeds is the same as the Judge. If you say, Such will be the condition of all souls in that trial; I shall reply,[3] To one only was it said, Thou art Peter, &c. And further, that the dignity of that see has been made venerable to the whole world by the voice of holy pontiffs, when all

---

[1] Hefele, ii. 625-30; Röhrbacher, viii. 463.
[2] Mansi, viii. 284, *The libellus apologeticus*, pp. 274-290.
[3] Replicabo, uni dictum, Tu es Petrus, &c., et rursus sanctorum voce pontificum dignitatem ejus sedis factam toto orbe venerabilem, dum illi quicquid fidelium est ubique submittitur, dum totius corporis caput esse designatur. —Mansi, viii. 284.

the faithful in every part are made subject to it, and it is marked out as the head of the whole body."

From the whole of this history we deduce the fact, that the enmity of the eastern emperor was able by bribing a party at Rome to stir up a schism against the lawful Pope, which had for its result to call forth the witness of the Italian and the Gallic bishops respecting the singular prerogatives of the Holy See. They spoke in the person of Ennodius and Avitus. We have, in consequence, recorded for us in black and white the axiom which had been acted upon from the beginning, " the First See is judged by no one ".

Let us see on the contrary what the same emperor was not only willing but able to do in the city which had succeeded to Rome as the capital of the empire, in which Anastasius reigned alone.

In the year 496, Anastasius had found himself able, as we have seen, to depose, by help of the resident council, Euphemius of Constantinople. As his successor was chosen Macedonius, sister's son of the former bishop, Gennadius, and like him of gentle spirit, " a holy man,[1] the champion of the orthodox ".[2] However much the opinion was then spread in the East that a successor might rightfully be appointed to a bishop forcibly expelled from his see, if otherwise the Church would be deprived of its pastor—an opinion which Pope Gelasius very decidedly censured—Macedonius II. felt very keenly the unlawfulness of his appointment. When the deposed Euphemius asked of him a safe

[1] The narrative from Photius, i. 134. [2] Ephrem, v. 9759.

conduct for his journey into banishment, and Macedonius received authority to grant it, he went into the baptistry to give it, but caused his archdeacon first to remove his omophorion, and appeared in the garb of a simple priest to give his predecessor a sum of money collected for him. He was much praised for this. Yet Macedonius had to subscribe the Henotikon. Hence he experienced a strong opposition from the monks, who, in their resolute maintenance of the Council of Chalcedon, declined communion with him; so the nuns also. Macedonius sought to gain them by holding a council in 497 or 498, which condemned the Eutycheans and expressed assent to the Council of Chalcedon.

Macedonius was by no means inclined to give up the lately won privileges of his see as to the ordination of the Exarch of Cappadocian Cæsarea, but he would willingly have restored peace with Rome, and have accepted the invitation from Rome to celebrate with special splendour the feast day of St. Peter and St. Paul. The emperor would not let him send a synodical letter to Rome.

Macedonius could not be induced by threat or promise of the emperor to give up to him the paper in which at his coronation by Euphemius he had promised to maintain the Council of Chalcedon. The emperor, after concluding peace with the Persians, more and more favoured the Eutycheans, and seemed resolved either to bend or to break Macedonius. The people were so embittered against Anastasius that he did not venture to appear without his life-guards even at a religious solemnity, and this became from that time a rule which marks the sinking moral influence of the emperors. The suspicion of the people against Anastasius was increased because his mother was a Manichean, his uncle, Clearchus, devoted to the Arians, and he kept in his palace Manichean pictures by a Syropersian artist. The Monophysite party had at the time two very skilful leaders, the monk Severus from Pisidia and the Persian Xenaias. Xenaias had been made bishop of Hierapolis by Peter the Fuller, was in fierce conflict with Flavian, patriarch of Antioch, and raised almost all Syria against him. He carried the flame of discord even to Constantinople. There a certain fanatic, Ascholius, tried to murder Macedonius, who pardoned him and bestowed on him a monthly pension. Presently large troops of monks came under Severus to Constantinople, bent upon ruining Macedonius. The state of parties became still more threatening. Macedonius showed still greater energy; he declared that he would only hold communion with the patriarch of Alexandria and the party of Severus if they would recognise the Council of Chalcedon as mother and teacher. But Anastasius, bribed by the Alexandrian patriarch John II. with two thousand pounds of gold, required that he should anathematise this council. To this Macedonius answered that this could not be done except in an œcumenical council presided over by the bishop of Rome. The emperor in his wrath violated the right of sanctuary in the Catholic churches and bestowed it on heretical churches. The Eutycheans supplied with money broke out against the Catholics.

They had sung their addition to the Trisagion on a Sunday in the Church of St. Michael within the palace. They tried to do it the next Sunday in the cathedral, upon which a fierce tumult broke out, and they were mishandled and driven out by the people. Now the party of Severus, favoured by the emperor and many officials, broke out into loud abuse of Macedonius. Thereupon the faithful part of his flock rose for their bishop, and the streets rung with the cry, "It is the time of martyrdom; let no man forsake his father". Anastasius was declared a Manichean and unfit to rule. The emperor was frightened; he shut the doors of his palace and prepared for flight. He had sworn never again to admit the patriarch to his presence, but in his perplexity sent for him. On his way Macedonius was received with loud acclaim, "Our father is with us," in which the life-guards joined. He boldly reproved the emperor as enemy of the Church; but the emperor's hypocritical excuses pacified the patriarch. When the danger was passed by Anastasius pursued fresh intrigues. He required Macedonius to subscribe a formula in which the Council of Chalcedon was passed over. Macedonius would seem to have been deceived, but afterwards insisted publicly before the monks on his adherence to its decrees. Then Anastasius tried again to depose him. All possible calumnies were spread against him—immorality, Nestorianism, falsification of the Bible; all failed. Then the emperor demanded the delivering up of the original acts of Chalcedon, which the patriarch steadily refused. Macedonius had sealed them up and placed them on the altar under God's protection; but the emperor had them taken away by the eunuch Kalapodius, economus of the cathedral, and then burnt. After this he imprisoned and banished a number of the patriarch's friends and relations; then he had the patriarch seized in the night, deported from the capital to Chalcedon, and thence to Euchaites in Paphlagonia, to which place he had also banished Euphemius. Macedonius lived some years after his exile. He died at Gangræ about 516, and was immediately counted among the saints of the eastern Church.

It cost Anastasius fifteen years to depose Macedonius, that is, from 496 to 511, and this was the way he accomplished it. Thus he succeeded in overthrowing two bishops of his capital—Euphemius and Macedonius—neither of whom lived or died in communion with Rome, because, though virtuous and orthodox in the main, they would not surrender the memory of Acacius. They had, moreover, one grievous blot on their conduct as bishops. They submitted themselves to subscribe an imperial statement of doctrine and to permit its imposition on others. This was a use of despotism in the eastern Church introduced by the insurgent Basiliscus, carried out first by Zeno and then by Anastasius, tending to the ruin both of doctrine and discipline. During the whole reign of Anastasius the patriarchal sees of Alexandria and Antioch, which had built up the eastern Church in the first three centuries, which Rome acknowledged as truly patriarchal under Pope Gelasius in 496, and the new sees which claimed to be patriarchal,

Constantinople and Jerusalem, were in a state of the greatest confusion, a prey to heresy, party spirit, violence of every kind. Anastasius was able to disturb Pope Symmachus during the first half of his pontificate by fostering a schism among his clergy, with the result that he brought out the recognition of the Pope's privilege not to be judged by his inferiors. But he was enabled to depose two bishops of the imperial see, his own patriarchs, blameless in their personal life, orthodox in their doctrine, longing for reunion with Rome, yet stained by their fatal surrender of their spiritual independence, subscription to the emperor's imposition of doctrine. They were not acknowledged by St. Peter's See, and they fell before the emperor.

In the last years of this emperor, the churches of the eastern empire were involved in the greatest disorders and sufferings. He had thrown aside altogether the mask of Catholic: he filled the patriarchal sees with the fiercest heretics. Flavian was driven from Antioch, Elias from Jerusalem. Timotheus, a man of bad character, had been put by him into the see of Constantinople. In this extremity of misery and confusion, the eastern Church addressed Pope Symmachus in 512.[1]

"We venture to address you, not for the loss of one sheep or one drachma, but for the salvation of three parts of the world, redeemed not by corruptible silver or gold, but by the precious blood of the Lamb of God, as the blessed prince of the glorious Apostles taught, whose chair the Good Shepherd, Christ, has entrusted

[1] Ecclesia orientalis ad Symmachum episcopum Romanum.—Mansi, viii. 221-6.

to your beatitude. Therefore, as an affectionate father for his children, seeing with spiritual eyes how we are perishing in the prevarication of our father Acacius, delay not, sleep not, but hasten to deliver us, since not in binding only but in loosing those long bound the power has been given to thee; for you know the mind of Christ who are daily taught by your sacred teacher Peter to feed Christ's sheep entrusted to you through the whole habitable world, collected not by force, but by choice, and with the great doctor Paul cry to us your subjects 'not because we exercise dominion over your faith, but we are helpers in your joy'. 'Hasten then to help that east from which the Saviour sent to you the two great lights of day, Peter and Paul, to illuminate the whole world.'" They call upon him as the true physician; they disclose to him the ulcerous sores with which the whole body of the eastern Church is covered; and they finish by directing to him a confession of faith, rejecting the two opposite heresies of Nestorius and Eutyches. They remind him of the holy Pope Leo, now among the saints, and conjure him to save them now in their souls as Leo saved bodies from Attila.

But yet it was not given to Pope Symmachus to put an end to this confusion. He sat during fifteen years and eight months, dying on the 9th July, 514. The schism raised by the Greek emperor was at an end; and seven days after his decease the deacon Hormisdas was elected with the full consent of all. In the meantime the state of the East had gone on from bad to worse. Anastasius, by writing and by oath, had pledged himself at his coronation to maintain the Catholic faith and the Council of Chalcedon. Instead he had persecuted Catholics, banished their bishops, by his falsehood and tyranny sown discord everywhere. At last one of his own generals, Vitalian, rose against him. After a long silence he once more betook himself to the Pope. In January, 518, he wrote to the new Pope, Hormisdas, " that the opinion spread abroad of his goodness led him to apply to his fatherly affection to ask of him the offices which our God and Saviour taught the holy Apostles by mouth, and especially St. Peter, whom He made the strength [1] of His Church". He asked, therefore, "his apostolate by holding a council to become a mediator by whom unity might be restored to the churches," and proposed that a general council should be held at Heraclea, the old metropolis of Thrace.

Hormisdas, after maturely considering the whole state of things, sent a legation of five persons to the emperor at Constantinople—the bishops Ennodius of Pavia, Fortunatus of Catania, the priest Venantius, the deacon Vitalis, and the notary Hilarius—with the most detailed instructions how to act. The intent was to test the emperor's sincerity—a foresight which after events completely justified. This instruction is said to be the earliest of the kind which has come down to us. Since nothing can so vividly represent the position of the Holy

---

[1] In qua fortitudinem Ecclesiæ suæ constituit. Epistola Anastasii ad Hormesdam pontificem.—Mansi, viii. 384.

See as the words used by it on a great occasion at the very moment when it took place, I give a translation of it. In reading this it should be remembered that these are the words of a Pope living in captivity under an Arian and barbaric sovereign, who had taken possession of Italy about twenty years before, and had sought for and accepted the royal title from this very emperor. Further, that with the exception of the Frankish kingdom, in which Clovis had died four years before, all the West was in possession of Arian rulers, who were also of barbaric descent. The Pope speaks in the naked power of his " apostolate ". The commission which he gave to his legates was this :[1]

"When, by God's help and the prayers of the Apostles, you come into the country of the Greeks, if bishops choose to meet you receive them with all due respect. If they propose a night-lodging for you do not refuse, that laymen may not suppose you will hold no union with them. But if they invite you to eat with them, courteously excuse yourselves, saying, Pray that we may first be joined at the Mystical Table, and then this will be more agreeable to us. Do not, however receive provision or things of that kind, except carriage, if need be, but excuse yourselves, saying that you have everything, and that you hope that they will give you their hearts, in which abide all gifts, charity and unity, which make up the joy of religion.

"So, when you reach Constantinople, go wherever the emperor appoints ; and before you see him, let no

[1] Mansi, viii. 389-393.

one approach you, save such as are sent by him. But when you have seen the emperor, if any orthodox persons of our own communion, or with a zeal for unity, desire to see you, admit them with all caution. Perhaps you may learn from them the state of things.

"When you have an audience of the emperor, present your letters with these words: 'Your Father greets you, daily intreating God, and commending your kingdom to the intercession of the holy Apostles Peter and Paul, that God who has given you such a desire that you should send a mission in the cause of the Church and consult his holiness, may bring your wish to full completion'.

"Should the emperor wish, before he receives your papers, to learn the scope of your mission, use these words: 'Be pleased to receive our papers'. If he answer, 'What do they contain?' reply, 'They contain greeting to your piety, and thanks to God for learning your anxiety for the Church's unity. Read and you will see this.' And enter absolutely into nothing before the letters have been received and read. When they have been received and read, add: 'He has also written to your servant Vitalian, who wrote that he had received permission from your piety to send a deputation of his own to the holy Pope, your Father. But as it was just to direct these first to your majesty, he has done so; that by your command and order, if God please, we may bear to him the letters which we have brought.'

"If the emperor ask for our letters to Vitalian,

answer thus: 'The holy Pope, your Father, has not so enjoined on us; and without his command we can do nothing. But that you may know the straightforwardness of the letters, that they have nothing but entreaties to your piety, to give your mind to the unity of the Church, assign to us some one in whose presence these letters may be read to Vitalian.' But if the emperor require to read them himself, you will answer that you have already intimated not such to be the command of the holy Pope. If he say, 'They may have also other charges,' reply, 'Our conscience forbids. That is not our custom. We come in God's cause. Should we sin against Him? The holy Pope's mission is straightforward; his request and his prayers known to all: that the constitutions of the fathers may not be broken; that heretics be removed from the churches. Beyond that our mission contains nothing.'

"If he say, 'For this purpose I have invited the Pope to a council, that if there be any doubt, it may be removed,' answer, 'We thank God, and your piety, that you are so minded, that all may receive what was ordered by the fathers. For then may there be a true and holy unity among the churches of Christ, if, by God's help, you choose to preserve what your predecessors Marcian and Leo maintained.' If he say, 'What mean you by that?' answer, 'That the Council of Chalcedon, and the letters of Pope St. Leo, written against the heretics Nestorius, and Eutyches, and Dioscorus, may be entirely kept'. If he say, 'We received and we hold the Council of Chalcedon, and the letters of Pope Leo,' do you then return thanks, kiss his breast, and say, 'Now we know that God is gracious to you, when you hasten to do this, for that is the Catholic faith which the Apostles preached, without which no one can be orthodox. All bishops must hold to this and preach it.'

"If he say, 'The bishops are orthodox; they do not depart from the constitutions of the fathers,' answer, 'If the constitutions of the fathers are kept, and what was decreed in the Council of Chalcedon is in no respect broken, how is there such discord in the churches of your land? Why do not the bishops of the East agree?' If he say, 'The bishops were quiet; there was no disunion among them. The holy Pope's predecessor stirred up their minds with his letters, and made this confusion;' answer, 'The letters of Symmachus, of holy memory, are in our hands. If, besides, what your piety says, that is, "I follow the Council of Chalcedon, I receive the letters of Pope Leo," they contain nothing except the exhortation to maintain this, how is it true that confusion has been produced by them? But if that is contained in the letters which both your Father hopes and your piety agrees to, what has he done? What is there in him blameworthy?' add your prayers and tears, entreat him, 'Let your imperial majesty consider God; put before your eyes his future judgment. The holy fathers who made these rules followed the faith of the blessed Apostle, on which the Church of Christ is built.'

"If the emperor say, 'I receive the Council of

Chalcedon, and I embrace the letters of Pope Leo, enter then into communion with me,' answer, ' In what order is that to take place ? We do not avoid your piety, so declaring, since we know that you fear God, and rejoice that you are pleased to keep the constitutions of the fathers. We therefore confidently entreat you that the Church may return through you to unity. Let all the bishops learn your will, and that you keep the Council of Chalcedon, and the letters of Pope Leo, and the apostolical constitutions.' If he say, ' In what order is that to take place ?' recur again, humbly, to entreaties, saying, ' Your Father has written to all the bishops. Join, herewith, your mandates to the effect that you maintain what the Apostolic See proclaims, and then let the orthodox not be separated from the unity of the Apostolic See, and the opponents will be made known. After that, your Father is even prepared, if need be, to be present himself, and, preserving the constitutions of the fathers, to deny nothing which is expedient for the Church's integrity.'

"If the emperor say, ' Well, in the meantime accept the bishop of my city,' again beseech humbly, ' Imperial majesty, we have come with God's help in the hope of support on your part to make peace and restore tranquillity in your city. There is question here about two persons. The matter runs its proper course. First, let all the bishops be so ordered as to form one Catholic communion ; next, the cause of those persons, or of any others who may be at a distance from their churches, can be specially considered.' If the emperor say, ' You are speaking of Macedonius ; I see your subtlety. He is a heretic ; he cannot possibly be recalled,' answer, ' Imperial majesty, we name no one personally ; we speak rather in favour of your mind and opinion, that inquiry may be made, and, if he is heretical, a juridical sentence passed, that he may not be said to be unjustly deposed, being reputed orthodox'.

"If the emperor should say, ' The bishop of this city consents to the Council of Chalcedon and the letters of Pope Leo,' answer, 'If he do so it will help him the more when his cause is examined ; and since you have allowed your servant Vitalian to treat with the Pope, if he hoped for a good result on these matters, so let it be'. If the emperor say, 'Should my city remain without a bishop, is it your desire that where I am there should be no bishop?' reply, 'We said before there was a question about two persons in this city. As to the canons, we have already suggested that to break the canons is to sin against religion. There are many remedies by which your piety may not remain without communion, and the full judicial form may be preserved.' If he say, 'What are those forms?' reply, 'Not newly invented by us. The question as to other bishops may be suspended, and meanwhile a person who agrees with the confession of your piety and with the constitutions of the Apostolic See until the issue of the trial may hold the place of the bishop of Constantinople, if by God's help the bishops are willing to be in accordance with the Apostolic See. You have in the records of the Church the terms of the profession which they have to make.'

"But if petitions be presented to you against other Catholic bishops, especially against those who shamelessly anathematise the Council of Chalcedon, and do not receive the letters of Pope St. Leo, take those petitions, but reserve the cause to the judgment of the Apostolic See, that you may give them a hope of being heard, and yet reserve the authority due to us. If, however, the emperor promise to do everything if we will grant our presence, urge in every way that his mandate first be sent to the bishops through the provinces, which one of you shall accompany, so that all may know that he keeps the Council of Chalcedon and the letters of Pope St. Leo. Then write to us that we prepare to come.

"It is, moreover, the custom to present all bishops to the emperor through the bishop of Constantinople. If their skilful management so devise in recognising your legation that you see the emperor in the company of Timotheus, who appears now to govern the church of Constantinople, if you learn before your presentation that this is so contrived, say, 'The Father of your piety has so commanded and enjoined us that we should see your majesty without any bishop'. So remain until this custom be altered.

"If an absolute refusal be given, or if it is so contrived that before you have an audience you are suddenly put with Timotheus, say, 'Let your piety grant us a private audience to set forth the causes for which we have been sent'. If he say, 'Speak before him,' answer, 'We do no offence, but our legation also contains his person, and he cannot be present at our communications'. And on no account enter into anything in his presence; but when he has gone out produce the text of your mission."

The exact conditions which the legates carried to the emperor were these: "The Council of Chalcedon and the letters of Pope St. Leo to be kept. The emperor, in token of his agreement, to send an imperial letter to all the bishops signifying that he so believes and will so maintain. The bishops also to express their agreement in Church in presence of the Christian people that they embrace the holy faith of Chalcedon and the letters of Pope St. Leo, which he wrote against the heretics, Nestorius, Eutyches, and Dioscorus, also against their followers, Timotheus Ailouros, Peter, or those similarly guilty, likewise anathematising Acacius, formerly bishop of Constantinople, and also Peter of Antioch, with their associates. Writing thus with their own hand in presence of chosen men of repute, they will follow the formulary which we have issued by our notary.

"Those who have been banished in the Church's cause are to be recalled for the hearing of the Apostolic See, that a trial and true examination may be held. Their cause to be reserved entire.

"If any holding communion with the sacred Apostolic See, preaching and following the Catholic faith, have been driven away, or kept in banishment, these, it is just, to be first of all recalled.

"Moreover, the injunction we have laid upon the

legates, that if memorials be presented to them against bishops who have persecuted Catholics, their judgment be reserved to the Apostolic See, that in their case the constitutions of the fathers be maintained, by which all may be edified."

Anastasius[1] tried again the old arts. He made a bid of everything to gain the legates. He seemed ready to accept everything save the demand regarding Acacius, which he was bound to reject on account of the Byzantine people. Both to the legates on their return to Rome, and to two officers of his court whom he sent to Rome, he gave honourable letters for the Pope, whom he invited to be present at the projected council, and endeavoured to satisfy fully by an orthodox profession of faith wherein he expressly recognised the Council of Chalcedon. One only point, he said, whatever might be his personal feeling, he could not concede, that regarding Acacius, since otherwise the living would be driven out of the Church for the dead, and great disturbances and blood-shedding would be inevitable. He left it to the Pope's consideration. He also wrote to the Roman senate to use its influence for the restoration of peace to the Church, as well with the Pope as with king Theodorick, "to whom," said the emperor, "the power and charge of governing you have been committed". It may be added that Theodorick favoured, as far as he could, the restoration of peace.

Pope Hormisdas, in his answer, praised the zeal made show of by the emperor, and wished that his deeds

---

[1] Photius, i. 143-5, translated.

would correspond to his words. He could not contain his astonishment that the promised embassy was so long in coming, and that the emperor instead of sending bishops to him, sent two laymen of his court, in whom he soon recognised Monophysites, who tried to gain him in their favour. In a letter to St. Avitus and the bishops of his province, he discloses the judgment which he had formed. "As to the Greeks, they speak peace with their mouth, but carry it not in their hearts; their words are just, not their actions; they pretend to wish what their deeds deny; what they professed, they neglect; and pursue the conduct which they condemned."[1] Still he resolved to send a new embassy to Constantinople in 517, at the head of which he put the bishops Ennodius and Peregrinus. He gave them letters to the emperor, the patriarch Timotheus, the clergy and people of Constantinople.

Anastasius had endeavoured to delay the whole thing, and to deceive the orthodox until he found himself strong again, and was no longer in danger from Vitalian. To bribe the people, he gave the church of Constantinople seventy pounds' weight of gold for masses for the dead. With regard to the treatment of Acacius, he had the majority on his side, who were not easily brought to condemn him. Here, also, he had a pretext to break off impending agreements. When his wife Ariadne died, he showed himself still less inclined to peace. She had been devoted to Macedonius, and often interceded for the orthodox. As soon as he thought himself quite

---

[1] *Ep. x. ad Avitum Viennensem.* Mansi, viii. 410.

secure, he not only altered his behaviour and language to the Roman See, but, in the words of the Greek historian, about 200 bishops who had come to Heraclea from various parts had to separate without doing anything, "having been deluded by the lawless emperor and Timotheus, bishop of Constantinople".[1] The Pope's legates he tried to corrupt; when that did not succeed, he dismissed them in disgrace, and sent the Pope an insolent letter, in which he said he desisted from any requests to him, as reason forbade to throw away prayers on those who would listen to nothing, and while he might submit to injuries, he would not endure commands. Thereupon broke out a great persecution against Catholics, which the Archimandrites of the second Syria report to Hormisdas.

In a supplication signed by more than two hundred, they address him:[2] "Most blessed Father, we beseech you, arise; have compassion on the mangled body, for you are the head of all. Come to save us. Imitate our Lord, who came from heaven on earth to seek out the strayed sheep. Remember Peter, prince of the Apostles, whose See you adorn, and Paul, the vessel of election, for they went about enlightening the earth. The flock goes out to meet you, the true shepherd and teacher, to whom the care of all the sheep is committed, as the Lord says, 'My sheep hear My voice'. Most holy, despise us not, who are daily wounded by wild beasts." All that the Roman See had gained was that the orthodox bishops and many conspicuous easterns

[1] Theophanes, p. 248.   [2] Mansi, viii. 425.

attached themselves to it, and the formulary binding them to obedience to the decisions of the Roman See found very many subscribers. The empire was in the greatest confusion when Anastasius died suddenly in the year 518, hated by the majority of his people, as perjured, heretical, and rapacious. Just before him died the heretical patriarchs, John II. of Alexandria and Timotheus of Constantinople.

Then suddenly,[1] as in the third century the Illyrian emperors saved the dissolving empire, another peasant, who in long and honourable service had risen to the rank of general, and was respected by all men as a virtuous man and a good Catholic, was called to take up that eastern crown of Constantine, which Zeno and Anastasius had soiled with the iniquities and perfidies of forty years.

At Bederiana, on the borders of Thrace and Illyria, there had lived three young men, Zimarchus, Ditybiotus, and Justin. Under pressure of misfortune they deserted the plough, and sought a livelihood elsewhere. They started on foot, their clothes packed on their backs, no money in their purses, with a loaf in their knapsacks. They came to Byzantium and enlisted. Twenty years of age and well grown, they attracted the notice of the emperor Leo I.: he enrolled them among his life-guards. Justin served as captain in the Isaurian war. For some unknown fault he was condemned to death by his general, and the next day was to be executed. The general, says Procopius, was

[1] German Röhrbacher, viii. 532, book 43, 81, mostly followed.

changed by a vision which he saw that night. Under Anastasius, Justin rose to the rank of senator, patrician, and commander of the imperial guard. On the death of Anastasius, the eunuch Amantius, who was lord chamberlain, and had been up to that time all powerful, sent for Justin, and gave him great sums of money to get the voice of the soldiers and the people, for a creature of his own, named Theocritus, in whose name he intended to rule. Justin distributed the money in his own name, and on the 9th July was proclaimed emperor by army and people. He was sixty-eight years old, and, if Procopius may be believed, could not even write his own name, at least in Latin. But he was of long experience, and admirable in the management of affairs. His wife was named Lupicina, of barbarian birth. Justin, in the first year of his service, had bought her as a slave, and married her. When he became emperor he crowned her as empress, and with the applause of the people gave her the name of Euphemia. He had a nephew born at Tauresium, a village of Dardania, near Bederiana. He was called Uprauda in his own land; his father was Istock, his mother Vigleniza. The Romans changed these Teuton names to Justinian, Sabbatius, and Vigilantia. Uprauda, the Upright, was the future emperor Justinian.

The accession of Justin was received with universal joy; and the new emperor at once sent a high officer, Gratus, count of the sacred consistory, to announce it to Pope Hormisdas, with a letter in which he said that "John, who had succeeded as bishop of Constantinople, and the other bishops assembled there from various regions, having written to your Holiness for the unity of the churches, have earnestly besought us also to address our imperial letters to your Beatitude. We entreat you, then, to assist the desires of these most reverend prelates, and by your prayers to render favourable the divine majesty to us and the commonwealth, the government of which has been entrusted to us by God."[1]

The count Justinian also wrote to Pope Hormisdas that "the divine mercy, regarding the sorrows of the human race, had at length brought about this time of desire. Thus I am free to write to your apostolate, our Lord, the emperor, desiring to restore the churches to unity. A great part has been already done. It only requires to obtain the consent of your Beatitude respecting the name of Acacius. For this reason his majesty has sent to you my most particular friend Gratus, a man of the highest rank, that you might condescend to come to Constantinople for the restoration of concord, or at least hasten to send bishops hither, for the whole world in our parts is impatient for the restoration of unity."[2]

The result was that Pope Hormisdas held a council at Rome in 518, at which all that had been done by his predecessors, the Popes Simplicius, Felix, Gelasius, and Symmachus, was carefully reviewed, and all present decreed that the eastern Church should be received into communion with the Apostolic See, if they condemned

---

[1] Mansi, viii. 435. [2] Mansi, viii. 438.

the schismatic Acacius, entirely effacing his name, and also expunged from the diptychs Euphemius and Macedonius, as involved in the same guilt of schism. And a pontifical legation was then named to carry out the desire of the council, and they bore with them an instruction, from which they might not depart by a hair's-breadth.[1]

The Pope wrote letters to the emperor, to the empress, to the count Justinian, especially to the bishop of Constantinople, recommending his legates, and exhorting the bishop to complete the work which was begun by condemning Acacius and his followers; also to the archdeacon Theodosius and the clergy of Constantinople.[2] He points out especially that he wants nothing new, or unusual, or improper, for Christian antiquity had ever avoided those who had associated with persons condemned; whoever teaches what Rome teaches, must also condemn what Rome condemns; whoever honours what the Pope honours, must likewise detest what he detests. A perfect peace admits of no division. The worship of one and the same God can only hold its truth in the unity of confession which embodies the belief.

The papal legates were received honourably on their journey, and found the bishops in general disposed to sign the formulary issued by the Pope. In March, 519, they came to Constantinople, where they found the greatest readiness. The patriarch John took the

[1] Mansi, viii. 441. Indiculus quem acceperunt legati Apostolicæ Sedis. It much resembles the former one, given to the legates sent to Anastasius.

[2] Photius, i. 148.

formulary, and gave it the form of a letter, which seemed to him more honourable than a formulary such as those who had fallen would sign. He prefixed to the document which the Pope required to be subscribed the following preface:

"Brother most dear in Christ, when I received the letters of your Holiness, by the noble count Gratus, and now by the bishops Germanus and John, the deacons Felix and Dioscorus, the priest Blandus, I rejoiced at the spiritual charity of your Holiness, in bringing back the unity of God's most sacred churches, according to the ancient tradition of the fathers, and in hastening to reject those who tear to pieces Christ's reasonable flock. Be then assured that, as I have written to you, I am in all things one with you in the truth. All those rejected by you as heretics I also reject for the love of peace. For I accept as one the most holy churches of God, yours of elder, and this of new Rome; yours the See of the Apostle Peter, and this of the imperial city, I define to be one. I assent to all the acts of the four holy councils—that is, of Nicæa, Constantinople, Ephesus, and Chalcedon—done for the confirmation of the faith and the state of the Church, and suffer nothing of their good judgments to be shaken; but I know that those who have endeavoured to disturb a single iota of their decrees have fallen from the holy, universal, and apostolical Church; and using plainly your own right words, I declare by this present writing,"[1] &c.

This is the preface given to his letter by the patriarch

[1] Mansi, viii. 451.

John; he then adds the formulary issued by the Pope from his council in Rome as the terms of restored communion between the East and West.

"The first condition of salvation is to maintain the rule of a right faith, and to deviate no whit from the tradition of the fathers; because the decree of our Lord Jesus Christ cannot be passed over, in which He says, 'Thou art Peter, and upon this rock I will build My Church'. These words are proved by their effect in deed, because the Catholic religion is ever kept inviolate in the Apostolic See. Desiring, therefore, not to fall from this faith, and following in all thing the constitutions of the fathers, we anathematise all heresies, but especially the heretic Nestorius, formerly bishop of Constantinople, condemned in the Council of Ephesus by Cœlestine, Pope of Rome, and the venerable Cyril, bishop of Alexandria; and together with him we anathematise Eutyches and Dioscorus, bishop of Alexandria, condemned in the holy Council of Chalcedon, which we follow and embrace with veneration, which followed the holy Nicene Council, and set forth the apostolic faith. To these we join Timotheus the parricide, surnamed Ailouros, and anathematise him, condemning in like manner Peter of Alexandria, his disciple and follower in all things; so also we anathematise Acacius, formerly bishop of Constantinople, who became their accomplice and follower, and those who persevere in communion and participation with them; for whoever embraces the communion of condemned persons shares their judgment. In like manner we condemn and anathematise Peter of Antioch, with all his followers. Hence we approve and embrace all the letters of St. Leo, Pope of Rome, which he wrote in the right faith. Therefore, as aforesaid, following in all things the Apostolic See, we preach all which it has decreed; and therefore I trust to be with you in that one communion which the Apostolic See proclaims, in which the solidity of the Christian religion rests entire and perfect,[1] promising that those who in future are severed from the communion of the Catholic Church, that is, who do not in all things agree with the Apostolic See, shall not have their names recited in the sacred mysteries. But if I attempt in aught to vary from this my profession, I declare that by my own condemnation I partake with those whom I have condemned. I have subscribed with my own hand to this profession, and directed it in writing to thee, Hormisdas, my holy and most blessed brother, and Pope of Great Rome, by the above-named venerable bishops, Germanus and John, the deacons Felix and Dioscorus, the priest Blandus."

The names of Acacius, Fravita, Euphemius, and Timotheus, four bishops of Constantinople, also of the emperors Zeno and Anastasius, who reigned from 474 to 518 (if we include a few months of Basiliscus), were erased from the diptychs in the presence of the legates. After that, at the instance of the emperor, the other bishops, the abbots, and the senate had signed the formulary, a solemn service was celebrated, to the great

---

[1] In qua est integra Christianæ religionis et perfecta soliditas.

joy of the people, in the Cathedral on Easter eve, the
24th March, to mark the act of reconciliation, and not
the least disturbance took place. The official narration[1]
of the five legates to Pope Hormisdas records the enthu-
siasm with which they were received at Constantinople.
"From the palace we went to the church with the vast
crowd. No one can believe the exultation of the people,
nor doubt that the Divine Hand was there, bestowing
such unity on the world. We signify to you that in
our presence the name of the anathematised prevaricator, Acacius, was struck out of the diptychs, as likewise that of the other bishops who followed him in
communion. So also the names of Anastasius and
Zeno. By your prayers peace was restored to the
minds of Christians: there is one soul, one joy, in the
whole Church; only the enemy of the human race,
crushed by the power of your prayer, is in mourning."

The emperor Justin wrote to Pope Hormisdas:

"Most religious Father, know that what we have so
long earnestly sought to effect is done. John, the
bishop of New Rome, together with his clergy, agrees
with you. The formulary which you ordered, which is
in agreement with the council of the most holy Fathers,
has been subscribed by him. In accordance with that
formulary, the mention at the divine mysteries of the
prevaricator Acacius, formerly bishop of this city, has
been forbidden for the future, as well as of the other
bishops who either first came against the apostolic con-

---

[1] Suggestio Germani et Joannis episcoporum, Felicis et Dioscori diaconorum, et Blandi presbyteri.—Mansi, viii. 453.

stitutions, or became successors of their error, and remained unrepentant to death. And since all our realm is to be admonished to imitate the example of the imperial city, we have directed everywhere our princely commands, so great is our desire to restore the peace of the Catholic faith to our commonwealth, to gain for my subjects the divine protection. For those whom the same realm contains, the same worship enlightens, what greater blessing can they have than to venerate with one mind laws of no human origin, but proceeding from the Divine Spirit? Let your Holiness pray that the divine gift of unity, so long laboured for by us, may be perpetually preserved."[1]

Thus history tells us that, in the year 484, Acacius, bishop of Constantinople, being condemned by Pope Felix, answered by striking the name of Pope Felix out of the diptychs, and that, in the year 519, the name of Acacius was erased from the diptychs in his own church; that his own successor not only gave up his memory, but, together with 2500 bishops,[2] signed a formulary which attributes to the Roman See the words of our Lord to St. Peter, which declares "that the Catholic religion is ever kept inviolate in the Apostolic See," "in which the solidity of the Christian religion rests entire and perfect," and which lays down the rule that whoever does not live and die in the communion of the Roman See has no claim to commemoration in the Church.

[1] Sacra imperatoris Justini ad Hormisdam.—Mansi, viii. 456.
[2] Photius, i. 149, who refers to the Deacon Rusticus, *Disputatio contra Acephalos*.

Let us now shortly review the facts which have passed under our notice since St. Leo returned from his interview with the pirate Genseric in the year 455.

In that fatal year the Theodosian house became extinct in the West so far as government was concerned. Valentinian's miserable widow, daughter of the eastern, wife of the western, emperor, during a short two months the prey of her husband's murderer, became with her daughters the captive of the Vandal freebooter, and saw the elder compelled to marry his son Hunnerich, the future persecutor of the Church. Twenty years succeed in which emperors are enthroned and pass like shadows, until the Herule general Odoacer, commanding for the time the Teuton mercenaries, deposes the last imperial phantom, Romulus Augustulus, and rules Rome and Italy with the title of Patricius. The western emperor is suppressed.

In 457, the Theodosian house becomes extinct in the East by the death of the emperor Marcian, before whom the heiress of the empire, St. Pulcheria, granddaughter of the great Thedosius, had died in 453. He was succeeded by Leo, a soldier of fortune, but an orthodox emperor, who supported St. Leo. The emperor Leo reigned until 474, and after a few months, in which his child grandson, Leo II., nominally reigned, the eastern crown was taken by Zeno and held till 491, with the exception of twenty months in which Basiliscus, a successful insurgent, was in possession. As Zeno had reigned in virtue of being husband of the princess Ariadne, daughter of Leo I., so Anastasius, in

491, in the words of the Greek chronicle, "succeeded to his wife and the empire," and he reigned twenty-seven years, to 518.

During this whole period, from the death of the emperor Leo I. in 474 to that of the emperor Anastasius in 518, the political state of the East and West was most perilous to the Church. In the East, the three sovereigns, Zeno, Basiliscus, and Anastasius, were unsound in their belief, treacherous in their action, scandalous in their life. The Popes addressed with honour, as the vice-gerents of divine power, men whom, as to their personal character, they must have loathed. Their government, moreover, was disastrous to their subjects—a tissue of insurrections, barbaric invasion, and devastation; at home, civil corruption of every kind.

In the West, Teuton conquerors had taken possession of the Roman empire. The Herule Odoacer had been put to death in 493 by the Ostrogoth Thedorick, who, like Odoacer before him, reigned with cognisance and approbation of the eastern emperor for thirty-three years. Both Odoacer and Theodorick were Arians; so also Genseric and his son Hunnerich, who ruled the former Roman provinces in Africa; so the Visigoths in southern France and Spain; so the Burgundians at Lyons. One conquering race only, that of the Franks, was not Arian, but pagan, until the conversion of Clovis, in 496, gave to the West one sovereign, Catholic and friendly to the Pope. We have seen in what terms Pope Anastasius welcomed his baptism. The population in the old Roman provinces which remained faithful to the Catholic religion was a portion of the old proprietors, such as had not been dispossessed by the successive confiscations and redistributions of land under the victorious northern invaders, and the poor, whether dwelling in cities or cultivating the soil. And these looked up everywhere to their several bishops for support and encouragement under every sort of trial. All men were sorted under two divisions in the vast regions for which Stilicho had fought and conquered in vain: the one division was Arian and Teuton, the other Catholic and Roman. And as the several Catholic people looked to their bishops, so all these bishops looked to the Pope; and St. Avitus expressed every bishop's strongest conviction when he said, writing in the name of them all, "In the case of other bishops, if there be any lapse it may be restored; but if the Pope of Rome is endangered, not one bishop, but the episcopate itself will seem to be shaken".

When the western emperor was suppressed the Pope became locally subject for about fourteen years to the Arian Odoacer, and then for a full generation to the Arian Theodorick. The latter soon found, by a calculation of interest, that the only way to rule Italy and the adjoining territories which his conquering arms had attached to Italy was by maintaining civil justice and equality among all his subjects. He took two of the noblest Romans, Boethius and Cassiodorus, for his friends and counsellors, and in the letters of the latter, from about the year 500 to the end of Theodorick's reign, we possess most valuable information as to the way in which Theodorick governed. Odoacer would seem likewise, during the years of his government until he was shut up in Ravenna, to have followed a like policy. But that the position of the Pope under Odoacer and Theodorick was one of great difficulty and delicacy no one can doubt. Gelasius speaks of his having had to resist Odoacer "by God's help, when he enjoined things not to be done".[1] And in 526 Pope John I. paid with his life, in the dungeon of Ravenna, the penalty for not having satisfied the Arian exactions of Theodorick in the eastern embassy imposed upon him.

I mention these things very summarily, having already given them with more or less detail, but I must needs recur to them because, in weighing the transactions which the schism of Acacius brought about, it is essential to bear in mind throughout the embarrassed and subject political situation in which all the Popes concerned with that schism found themselves.

Within seven years after the western emperor had been suppressed, and the overlordship of the East been acknowledged by the Roman senate as well as the Teuton conqueror, what happened?

A bishop of Constantinople, as able and popular as he was unscrupulous, had established a mental domination over the eastern emperor Zeno. He reigned in the utmost sacerdotal pomp at Constantinople; he beheld Old Rome sunk legally to the mere rank of a municipal

---

[1] Mansi, viii. 60.

city, and the See of St. Peter in it subject to an Arian of barbaric blood. He thought the time was come for the bishop of the imperial city to emancipate himself from the control of the Lateran Patriarcheium. Having gained great renown by his defence of the Council of Chalcedon against the usurper Basiliscus, having denounced at Rome the misdeeds and the heresy of the Eutychean who was elected by that party at Alexandria, and having so been high in the trust of Pope Simplicius, he turned against both Pope and Council. He set up two heretics as patriarchs—Peter the Stammerer, the very man he had denounced, at Alexandria, and Peter the Fuller at Antioch. He composed a doctrinal statement, called the "Form of Union," which, by the emperor's edict, was imposed on the eastern bishops. It was a scarcely-veiled Eutychean document. He called to his aid all the jealousy which Nova Roma felt for her elder sister, all the pride which she felt for the exaltation of her own bishop. If he succeeded in maintaining his own nominees in the two original patriarchates of the East, he succeeded at the same time in subjecting them to his own see. He crowned that series of encroachments which had advanced step by step since the 150 bishops of the purely eastern council held at Constantinople just a hundred years before set the exaltation of the imperial city on a false foundation. In fact, if this his enterprise succeeded, he obtained the realisation of the 28th canon, which Anatolius attempted to pass at Chalcedon, and which Pope Leo had overthrown. But most of all, both in the government of the Church and in the supreme magisterium, the determination of the Church's true doctrine, he deposed the successor of St. Peter, and but one single step remained, to which all his conduct implied the intention to proceed. For the logical basis of that conduct was the assertion that, as the bishop of Rome had been supreme when, and because, Rome was the capital of the empire, so when Constantinople had succeeded Rome as capital, her bishop also succeeded to the spiritual rights of the Primacy.

We may sum up the attempt of Acacius in a single word: the denial that the Pope had succeeded to the universal Pastorship of St. Peter.

This, then, was the point at issue, and when the western emperor was suppressed, and the overlordship of the eastern emperor acknowledged, the Pope was deprived of all temporal support, and left to meet the attack of Acacius in the naked power of his apostolate. From the year 483, when the deeds of Acacius led to his excommunication, followed by the schism, to its termination in 519, the Popes, being subjects of Arian sovereigns, who were likewise of barbaric descent, braved the whole civil power of the eastern emperors, as well as the whole ecclesiastical influence of the bishops of Constantinople. Not only were Zeno and Anastasius unorthodox, but likewise they were bent on increasing the influence of that bishop whom they nominated and controlled. The sovereigns of the East had been able, even by a simple practice of Byzantine etiquette, to put their own bishop in a position of determining influence over the whole eastern episcopate. For we learn from the instruction of Pope Hormisdas to his legates that it was the custom for every bishop to be presented to the emperor by the bishop of Constantinople. The Pope most strictly enjoins his legates not to submit to this. The effect of such a rule upon the eastern bishops who frequented the court of an absolute sovereign exhibits another cause of that perpetual growth which accrues to the bishop of the imperial city.

Every human power, every conjunction of circumstances, seemed to be against the Popes in this struggle. While the East was thus in hostile hands, under emperors who were either secretly or avowedly heretical, the West was under Arian domination. Italy was ruled from 493 to 526 by a man of great ability. Few rulers have surpassed Theodorick either in success as a warrior or in political skill. He had, further, enlaced the contemporary rulers in the various countries of the West in ties of relationship with himself. He had married Andefleda, sister of Clovis; he gave Theudigotha, one of his own daughters by a concubine, to Alaric of Toulouse, king of the Visigoths, and another, Ostrogotha, to Sigismond, king of the Burgundians, at Lyons. Even before he had conquered Odoacer, in 493, he was in strict alliance with the king of the Vandals in Africa, to whom he gave his sister Amalafrieda to wife, and her daughter Amalaberga to the king of the Thuringians. He solicited the royal title in 496 by an embassy to Anastasius, and the result of that embassy was that the chief man in it, Faustus, patrician and senator, when he returned to Rome, contrived to raise a schism in the clergy itself against Pope Symmachus. This schism was the greatest difficulty which the Pope in all this period encountered. Theodorick in political talent and warlike genius reminds historians of Charlemagne; but instead of having that monarch's faith, he was an Arian. His equal treatment of Arian and Catholic was a carefully thought-out policy; nor did he scruple at the very end of his career to sacrifice even the very life of the Pope to his political schemes. He favoured the senate of Rome in its corporate capacity; he favoured individual senators, but always as instruments of his own absolute rule, the key to which was to unite the use of the Roman mind in administration with the Gothic arm in action. When the end of the schism came, he had married his only child Amalasunta, the heiress of his kingdom, to Eutharic, who in the first year of the emperor Justin was consul of Rome with that prince, and nominated by him.

On what, then, did the Pope rely? On one thing only—that in the inmost conscience of the Church, in East and West, he was recognised as St. Peter's successor; that upon everyone who sat in the Apostolic See had descended the mighty inheritance, the charge which no man could execute except he were empowered by divine command and sustained by divine support. For as it required God to utter the words, "Upon this rock I will build My Church"; "If thou lovest Me, feed My sheep"; "Confirm thy brethren"; so it no less required God to enable any man to fulfil that charge. But how when it comes to a succession of men? How many families can show a continuous succession of three temporal rulers equally great? Can any family show four such? Can anyone calculate the power which maintains such a succession through centuries?

Here, after four full centuries, in that one belief the seven next successors of St. Leo—Hilarus, Simplicius, Felix, Gelasius, Anastasius, Symmachus, and Hormisdas —stood as one man. Their counsels did not vary. Their resolve was one. Their course was straight. In Leo's time the earth reeled beneath the tread of Attila, the city groaned beneath Genseric's hoof. And now three heretics—despots, and ignoble despots, if ever such there were—filled the sole imperial throne. Arians, closely connected by family ties and identical interests, divided the West among them. The seven Popes sat on at the Lateran in the palace which Constantine had given them, and said Mass in the church which he had built for them. Three of his degenerate successors tried every art against them and failed. During twenty years of this time, from 476 to 496, no ruler small or great acknowledged the Catholic faith. The East was Eutychean, the West Arian. At length St. Remigius baptised the Frankish chief as first-born of the Teuton race in the Catholic faith of the Holy Trinity, and the Pope at Rome gave utterance as a father to his joy. The end was that the schism was terminated on the part of the bishop, the heir of the seat and the ambition of Acacius, by the prince, by his nobles, among them the legislator who was to be Justinian, and by 2500 bishops throughout the East, acknowledging in distinct terms that one unique authority on which the Popes had rested throughout the contest. They declared solemnly, in celebrating the holiest mystery of the Christian faith, that the word of the Lord cannot be passed over, saying, "Thou art Peter, and upon this rock I will build My Church". They added that the course of five hundred years had exemplified the fact "that the solidity of the Christian religion rests entire and perfect in the Apostolic See". The rebellion of Acacius in 483 drew forth this confession from his successor, John II., in 519.

The seven successors of St. Leo stood as one man. No variation in their language or their conduct can be found. Not so the seven successors of Anatolius at Constantinople. That bishop, who had seen himself foiled by the vigour and sagacity of St. Leo at the Council of Chalcedon, lived afterwards on good terms with him, and died in 458, in his lifetime. He was succeeded by Gennadius, who, during the thirteen years of his episcopate, was faithful both to the creed which St. Leo had preserved and to the dignity of the Apostolic See. He was followed by Acacius, who occupied the see from 471 to 489. There was some quality in Acacius which gained the favour of princes. He had charmed at once the old emperor Leo I.; but Zeno, whose influence first made him bishop, afterwards followed all his teaching. He had also gained a renown for orthodoxy by refusing the attempt of Basiliscus to make the imperial will a rule of Church doctrine. It was when his stronger mind had mastered Zeno that he began the desperate attempt against the doctrine and discipline of the Apostolic See which has been our chief subject. But when he died in 489, his successor Fravita at once renounced the position which he had taken up by asking the recognition of Pope Felix and restoring his name in the diptychs. It is true that in his conduct he was double-dealing, and, while he sought for the Pope's recognition, parleyed with the heretical patriarch of Alexandria. But he died in three months, and was succeeded by Euphemius, who likewise repudiated the act of Acacius, and earnestly sought reconciliation with the Pope, while he was unwilling to fulfil the condition of it—that he should erase the name of Acacius from the dipytchs. The six years' episcopate of Euphemius was one long contest with the treachery and persecution of the emperor Anastasius, who at last, by help of the resident council, was able to depose him. He placed Macedonius in his stead, who again sought to be reconciled with the Pope, but only would not pay the price of renouncing the person, as he fully renounced the conduct, of Acacius. During fifteen years, from 496 to 511, as Euphemius had resisted the covert heresy of Anastasius, so did Macedonius, and, like him, he fell at last before the enmity of the emperor. Upon the deposition of Macedonius, the emperor obtained the election of Timotheus, who during seven years was his docile instrument. When he died in 518, the bishop John was elected, whose great desire was the restoration of unity, with the maintenance of the faith of Chalcedon. By side of the seven Popes succeeding St. Leo put the seven bishops of the emperor's city. We find two—the first and the last—Gennadius and John, blameless. The second, Acacius, author of all the evil in a schism of thirty-five years. The third, the fourth, and the fifth shrink from the deed of Acacius; and two of them are deposed by the emperor, while his people respect and cherish their memory. The sixth is a mere tool of the emperor.

Four eastern emperors occupy the sixty years from Marcian to Justin. Three of them are of the very worst which even Byzantium can show. Their reply to the appeal of the Pope to "the Christian prince and Roman emperor" was to betray the faith and sacrifice Rome to Arian occupation.

But when we turn from the bishops and emperors of the eastern capital to the seats of the ancient patriarchs, to the Alexandria of Athanasius and Cyril, to the Antioch of Ignatius, Chrysostom, and Eustathius, no words can express the division, the scandals, the excesses, which the Eutychean spirit, striving to overthrow the Council of Chalcedon, showed during those sixty years. With this spirit Acacius played to stir up the eastern jealousy against the Apostolic See of the West, and he found a most willing coadjutor in the eastern emperor, the more so because that See was no longer locally situated in his domain. The chance of Acacius lay throughout in the pride of that monarch who was become the sole inheritor of the Roman name, as Pope Felix reminded him, and who would fain see Nova Roma the centre of ecclesiastical rule, as it was become the head of the diminished empire. Anastasius, after Zeno, was still more swayed by these motives than his predecessor.

But here we touch the completeness of the success which followed the trust placed in their apostolate by the seven immediate successors of St. Leo. In proportion as Rome became in the temporal order a mere municipal city, the sacerdotal authority of its bishop came out into clearer light. Three times in the fifth century Rome was mercilessly sacked—in 410, in 455, in 472. Its senators were carried into slavery, its population diminished. The finishing stroke of its ignominy may be said to be the deposition, by a barbarian *condottiere*, of the poor boy whose name, repeating in connection the founder of the city with the founder of the empire, seemed to mock the mortal throes of the great mother. But this lessening of the secular city, so far from lessening the authority of the spiritual power, reveals to all men, believers or unbelievers, that the pontificate, whose seat is locally in the city, has a life not derived from the city. Rome's temporal fall exhibits in full the intangible spiritual character of the pontificate. If St. Peter had to any seemed to rule because he was seated on the pedestal of the Cæsarean empire, when that empire fell the Apostle alone remained to whom Christ gave the charge, whom

He invested with the "great mantle".[1] The bishop of the city in which an Arian Ostrogoth ruled supreme as to temporal things was acknowledged by the head of the empire, from whom the Ostrogoth derived his title, as the person in whom our Lord's word—the creative word which founds an empire as it makes a world—was accomplished, had been during five hundred years accomplished, would be for ever accomplished.[2]

The malice of Acacius largely led to this result. His attack was the prelude to the sifting of the Pope's prerogative during thirty-five years: its sifting by a rival at Constantinople, by the eastern bishops, by the eastern emperor, who had now also become the sole Roman emperor; and the sifting was followed by a full acknowledgment. Nothing but this hostile conduct would have afforded so indubitable a proof of the thing impugned. While the ancient patriarchates which had formed the substructure of the triple dais on which the Apostolic See rested were falling into irretrievable confusion, while the new State-made patriarch at Constantinople was trying to nominate and, if he could, to consecrate his elders and superiors at Alexandria and Antioch, who descended from Peter, the essential prerogative of the Apostolic See itself came forth into full light. The bishops at Alexandria, Antioch, Constantinople, Jerusalem, and every other city in the world would be great or small in influence according to the greatness or smallness of their city.

---

[1] Il granto manto, Dante.
[2] Quia in sede Apostolia inviolabilis *semper* Catholica custoditur religio.

If the city fell altogether, the see would fall. Its life was tied to the city. But it was not so with that pontificate on which the Church was built. There and there only the living power was given by Christ to a man: not local, nor limited, nor transitory. This was the great truth which the Acacian schism helped to establish in the minds of men, and which was proclaimed in that Nova Roma where Acacius had refused the judgment of Pope Felix, and had tried to put himself on an equality. As a result, in the terms of union which have been above recited, the action of Acacius has had the honour to condemn the rebellion of Photius three hundred years before it arose, and every other rebellion which has imitated that of Photius.

Nor must it be forgotten that it was the constancy of the Popes in these sixty years which alone prevented the prevailing of Eutychean doctrine in the East. Blent with that doctrine was the attempt of three emperors to substitute themselves as judges of doctrine for the Apostolic See and the bishops in union with it. At the moment when John Talaia[1] was expelled from Alexandria, the Monophysite heresy, espoused by Acacius and imposed by Zeno, would have triumphed, save for the Popes Simplicius and Felix. And it would have triumphed while the instrument of its triumph, the Henotikon, would have inflicted a deadly blow upon the government of the Church by taking away the independence of her teaching office. This struggle continued during the reign of Zeno; and Anastasius, as

[1] Hergenröther, *K. G.*, i. 333.

soon as he became emperor, used all the absolute power which he possessed to enforce the reception of the same document. Even Euphemius and Macedonius were obliged to sign it, and the sacrifice which they made in suffering deposition does not deliver their character of bishops from the stain of this weakness. We see in this period the first stadium traversed by the Greek Church in that descending course which, in another century, brought it to the ruin wrought by Mahomet.

On the other hand, the seven Popes kept the position of St. Leo—rather, they more than kept it, because, under outward circumstances so greatly altered for the worse, they both maintained his doctrine and justified his conduct. They insisted through the darkest times, under pressure of the greatest calamities, deprived of all temporal aid, that the person of Acacius should be solemnly removed from recognition as a bishop by the Church. They insisted, and it was done. The act of Acacius, if allowed to pass, would have carried into actual life the assertion of the canon which St. Leo had rejected: that the privileges of the Roman See were derived from the grant of the Fathers to Rome because it was the capital. The expunging of his name from the diptychs, with the solemn asseveration that the rank of the Holy See was derived from the gift of Christ, and that the Church's solidity as a fabric consisted in it, and equally the maintenance of the Catholic religion, established the contradictory of that 28th canon, and enforced for ever the subordination of the see which Acacius sought to exalt. At the same time it pointed out the distinction between the See of Peter and all other sees: the distinction that in the case of every other bishop the spiritual life of the bishop, as a ruler, is local and attached to his see. But the See of Peter is the generator of the episcopate, because of Peter ever living in his successor.

It may also be remarked that it is this overflowing life of Peter which invests titular bishops with the names of dead sees. Thus they sit as members of a General Council, verifying to the letter St. Cyprian's adage, that the episcopate is one, of which a part is held by each without division of the whole.

The submission of Constantinople in its bishop, its clergy, its emperor, its nobles, attested by the subscription of 2500 bishops throughout the East, is an event to which there can hardly be found a parallel. The submission was made to Pope Hormisdas when he was himself, as his predecessors for forty-three years had been, subject to an Arian ruler.[1] If there be in all history an act which can be called in a special sense an act of the undivided Church, it is this. It was made more than three hundred years before the schism of Photius. If the confession contained in this submission does not exhibit the mind of the Church, what form of words, what consent of will, can ever be shown to convey it? If those who subscribed this confession subscribed a falsehood, why pretend any longer to attribute authority to the Church? But it must be added, if their confession was the truth, why not obey it?

---

[1] See Photius, i. 149.

It is to be noted that this period of sixty years is full of events which caused the greatest suffering to the Popes, were unceasingly deplored by them, and resisted to the utmost of their power. The temporal condition of themselves, of the bishops, of their people in Italy, Africa, France, Spain, Illyricum, Britain, was most sad. The most vehement of persecutions desolated Africa. Again, there was the suppression of the western emperor, with the consequent subjection of the Apostolic See to the temporal government of the most hateful of heresies: the Oriental despotism of Zeno and Anastasius, continued for forty-four years, mixed with another heresy, and tending to destroy both faith and independence in the bishops subject to it. The Popes, as Romans, felt with the keenest sympathy the political degradation of Rome. Can any appeal be more touching than that which they made, and made in vain, to the "Christian king and Roman prince"? Out of all these things, whose natural consequences tended to extinguish their principate, came forth the most magnificent attestation to it which is to be found in the first five hundred years of the Christian religion.

CHAPTER IV.

JUSTINIAN.

The submission of the eastern empire and episcopate to Pope Hormisdas, in 519, is a memorable incident in the history of the Church. A large and marked part in it was taken by the man who for thirty-eight years was to rule the eastern empire, to expel the Goths from Italy, thus recovering the original seat of Roman power, and the Vandals from Africa, and so once more attach the great southern provinces, for so many ages the granary of Rome and Italy itself, to the existing Byzantine realm. Before, however, this was done, when, after the death of Theodorick, the Gothic kingdom still subsisted under his grandson Athalarick and his daughter Amalasunta, the emperor Justinian addressed to Pope John II., in the year 533, a letter from which I quote as follows. I preface that this letter was carried to the Pope by two imperial legates, the bishops Hypatius and Demetrius. It begins:[1] " Rendering honour to the Apostolic See and to your Holiness, whom we ever have revered, and do revere, as is befitting a father, we hasten to bring to the knowledge of your Holiness everything

---

[1] Mansi, viii. 795-99.

which concerns the state of the churches. For the existing unity of your Apostolic See, and the present undisturbed state of God's holy churches, has always been a thing which we have earnestly sought to maintain. And so we lost no time in subjecting and uniting all bishops of the whole eastern region[1] to the See of your Holiness. We have now, therefore, held it necessary that the points mooted, though they are clear and beyond doubt, and have been ever firmly maintained and proclaimed by all bishops according to the teaching of your Apostolic See, should be brought to the knowledge of your Holiness. For we do not allow that anything concerning the state of the churches, clear and undoubted though it be, when once mooted, should not be made known to your Holiness, who is the head of all the holy churches. For, as we said, in all things we hasten to increase the honour and authority of your See." He then proceeds to recite a creed which carefully condemns the errors of Nestorius on the one side, and Eutyches on the other, and acknowledges "the holy and glorious Virgin Mary to be properly and truly Mother of God". At the beginning of this creed he introduces the words: "All bishops of the holy and apostolic Church, and the most reverend archimandrites of the sacred monasteries, following your Holiness, and maintaining that state and unity of God's holy churches which they have from the Apostolic See of your Holiness, changing no wit of that ecclesiastical state which has

---

[1] This refers to the reunion of a great portion of the eastern Church, which had fallen a prey to the most manifold errors since the Council of Chalcedon. —Riffel, p. 543.

held and holds now, confess with one consent," &c. And he concludes with the words: "All bishops, therefore, following the doctrine of your Apostolic See, so believe, confess, and preach: for which we have hastened to bring this to the knowledge of your Holiness, by the bishops Hypatius and Demetrius; and we beg your fatherly affection, that by letters addressed to us, and to the bishop and patriarch, your brother, of this imperial city (since he on the same occasion wrote to your Holiness, being earnest in all things to follow the Apostolic See), you would make known to us that your Holiness receives all who make the above true confession. For so the love of all to you and the authority of your See will increase, and the unity of the holy churches with you will be preserved unbroken, when all bishops learn through you the sincere doctrine of your Holiness in what has been reported to you. But we beseech your Holiness to pray for us, and obtain for us the guardianship of God."

Pope John II. acknowledges this letter to "his most gracious son, Justinian Augustus". He highly celebrates the praises of "the most Christian prince," that "in your zeal for faith and charity, instructed in the Church's discipline, you preserve reverence to the See of Rome, and subject all things to it, and bring them to its unity, to the author of which, the first Apostle, the Lord's words were addressed, 'Feed My sheep': which both the rules of the Fathers and the statutes of emperors declare to be the head of all churches, and the reverential words of your Piety attest". The Pope adds:

"Your imperial words, brought by the bishops Hypatius and Demetrius, which have been agreed to by our brethren and fellow-bishops, being agreeable to apostolic doctrine, we by our authority confirm". "This, then, is your true faith; this all Fathers of blessed memory and prelates of the Roman Church, whom in all things we follow, this the Apostolic See has to this time preached and maintained unshaken." "And we beseech our God and Saviour Jesus Christ to preserve you long and peacefully in this true religion and unity, and veneration of the Apostolic See, whose principate you, as most Christian and pious, preserve in all things."

In the same year, 533, in which Justinian addressed to the Pope this remarkable recognition of the Roman Primacy, specifying that everything which concerns the whole Church should be brought before the Pope, though it might be already certain and in accordance with established usage, he gave his approval to that collection of laws called in Latin the *Digest* and in Greek the *Pandects*, which he had commissioned Tribonian and other great lawyers to draw up. Seventeen commissioners, having power given to them to alter, omit, and correct, selected by his command, out of nearly two thousand volumes, what they considered serviceable in the imperial laws and the decisions of great lawyers. It is a vast repertory of judicial cases in which Roman lawyers seek to apply the general rules of law and natural equity. It was the first attempt since the Twelve Tables to construct an independent centre of right as a whole,[1] and it was confirmed by the authority of the emperor on the 16th December, 533.

As in the whole course of the fifth century, so no less in the sixth, it is necessary to bear in mind the close interweaving of political with ecclesiastical facts. The force and bearing of the one only become intelligible when the others are weighed. In 519, under Pope Hormisdas, the schism of Acacius had collapsed, and the most emphatic acknowledgment of all which the Popes had claimed in the contest with him, and with the emperors Zeno and Anastasius, who favoured him, had taken place. Pope Hormisdas had been succeeded in 523 by Pope John I. Compelled by the king Theodorick to undertake an embassy to the emperor Justin, received at Byzantium with the highest honour as first Bishop of the Church, being also the first Pope who had visited the eastern capital, and crowned with gifts for the churches at Rome, he returned only to die in the dungeon of the Arian prince at Ravenna, in 526. In three months Theodorick had followed to the tomb his three victims — Symmachus, Boethius, and Pope John I. His death had well-nigh broken up the league of Teutonic Arian rulers against the Catholic faith, of which he had been the soul during the thirty-three years of his reign. Justinian had been taken by his uncle Justin as partner of his empire in April, 527, and crowned, together with his wife Theodora, on Easter Day. Four months later he succeeded his uncle in the

---

[1] Savigny, *Geschichte des römischen Rechts im Mittelalter*, 1834, i. 36. Quoted by Rump, ix. 72.

sole power. At the death of Theodorick, the innate weakness of the Gothic kingdom in Italy, which had been veiled by the personal ability of the sovereign, came to full light. The utter incompatibility between the savage Goth and the cultured Roman showed itself in the rejection of the queen Amalasunta, in the depriving her of her son, and his subsequent corruption and premature death, its result. It was shown also in the retirement of Cassiodorus from the place of counsellor and minister of the Gothic king. Upon the death of Pope John I., in 526, Theodorick had exercised his power in urging the Romans to select Felix for pope. For this permanent injury had been inflicted upon the liberty of the papal election by the foreign occupation of Italy. It began under Odoacer in 483, when the temporal ruler, being a foreigner and an Arian, for the first time sought to mix himself with the election. Twenty years after, under Pope Symmachus, the attempt of Odoacer had been condemned. But what the Herule and the Gothic ruler, both Arians, had begun, the Byzantine emperor, when he recovered possession of Rome, carried on, and the original freedom of election was subjected to the control of the eastern emperor for hundreds of years.

Pope Felix sat until 530, and was then succeeded by Bonifacius II., the son of a Goth; not, however, without a temporary schism, occasioned by the attempt of King Athalarick to exert the arbitrary power used by his grandfather Theodorick in the election. Pope John II. followed in 532. In this Pope's time Cassiodorus was made Prætorian prefect by King Athalarick, and wrote to the Pope as a son to his father: "Be careful to remind me what I am to do. I wish to deal rightly, though I am blamed. A sheep which desires to hear the voice of his shepherd is not so easily led astray; and if he has one who warns continually at his side, can scarcely be criminal. I am, indeed, judge in the palace, but shall not therefore cease to be your disciple. For we execute this office well when we do not in the least depart from your injunctions. Since, then, I wish to be guided by your counsels and supported by your prayers, you must show your hand when there is anything in me otherwise than would be desired. That chair which is the wonder of the whole world should carefully protect its own, since, though it is given to the whole world, yet it admits in you a special local love." [1]

The Pope, to whom the Prætorian prefect of Athalarick, the temporal sovereign, addressed this language, is John II., to whom Justinian, from Byzantium, spoke as a son, and whose primacy he acknowledged in terms so ample, before he became, by the conquest of Belisarius, the temporal lord of Rome; the year, also, before he reconquered Northern Africa by the sword of the same great general.

Justinian, with not less precision than former emperors, acknowledged all his life long the primacy of the Roman See. We need not exclude political motives from this acknowledgment, but we must allow to him

---

[1] *Ep.* xi. 2 : Sedes illa toto orbe mirabilis licet generalis mundo sit prædita.

the fullest conviction as to its legitimate authority. If now and then, under the impulse of passion or despotic humour, he seemed to disregard its rights, he soon strove again to obtain the Pope's assent to his measures. In his edict to his own patriarch Epiphanius, he declared expressly that he held himself bound accurately to inform the Pope, as head of all bishops, concerning the circumstances of his realm, especially since the Roman Church by its decisions in faith had overthrown the heresies which arose in the East.[1] The imperial theologian was very unwilling to give up the initiative in the determination of ecclesiastical questions; nevertheless, he acknowledged in the Bishop of Old Rome the superior judge without whose confirmation his own steps remained devoid of force and effect.[2]

The man who was born an Illyrian peasant, who was the leading spirit during the nine years' reign of another Illyrian peasant, his uncle, who succeeded him in 527, and ruled the greatest kingdom of the earth during thirty-eight years; to whom the bitter Vandal in Africa and the nobler Goth in Italy yielded up their equally ill-gotten prey; who became the great legislator of the Roman world, by the commission given to his chief lawyers to select and, after correction, tabulate the laws of the emperors his predecessors; to whom, in con-

---

[1] *Nov.* cxxxi. c. 2 : θεσπίζομεν τὸν ἁγιώτατον τῆς πρεσβυτέρας Ῥώμης πάπαν πρῶτον εἶναι πάντων τῶν ἱερέων. . . . τῇ γνώμῃ καὶ ὀρθῇ κρίσει τοῦ ἐκείνου σεβασμίου θρόνου κατηργήθησαν. *Nov.* ix. init. : Pontificatus apicem apud eam (Romam anteriorem) esse nemo est qui dubitet.—Photius, p. 156.

[2] Translated from Photius, p. 156.

sequence, the actual nations of Europe owe what was to them the fountain of universal right, demands a somewhat detailed account of his character, his purposes, and his actions. When the prince of the poets of Christendom, the only poet who has spoken in the name and with the voice of Christendom, meets his spirit under the guidance of Beatrice, the emperor utters words the truth of which all must feel:

> "Cæsar I was and am Justinian,
> Moved by the will of that Prime Love I feel
> I clear'd the encumbered laws from vain excess".[1]

It is in this character that Justinian lives for all history, and his name stands out among all Byzantine sovereigns with a lustre of its own. I have therefore first quoted the most definite words of the great legislator, spontaneously acknowledging the right of St. Peter's successor to know and to judge of all that concerns the Church's doctrine and practice. The acknowledgment of this right is the more to be marked because, when it was made by the eastern emperor, that successor was not his own subject. That he was the head of all the churches of the world, that he was so by descent from Peter, that in virtue of this headship and descent he had a right of supervision over everything which belonged to the Church in all the world—this is what Justinian avows, and this, moreover, is equally what the Pope claimed then as he claims now.

[1] "Cesare fui e son Giustiniano,
Che, per voler del primo amor ch' io sento,
Dentro alle leggi trassi il troppo e il vano."
—*Paradiso*, vi. 10.

Justinian ascended the eastern throne in August, 527, at about the age of forty-five. He would therefore have been born in 482. He was of somewhat more than middle height, of regular features, dark colour, of ample chest, serene and agreeable aspect. Through the care of his uncle he had had a good education, and had early learned to read and write. He was skilled in jurisprudence, architecture, music, and, moreover, in theology. His personal piety was remarkable. When he became emperor he bestowed all his private goods on churches, and ruled his house like a monastery. In Lent, his life approached that of a hermit in severity. He ate no bread; drank only water; for his nourishment he contented himself every other day with a portion of wild herbs, seasoned with salt and vinegar. We have sure testimony respecting his fasts and mortifications, since he has taken pains in his last laws, the *Novels*, to inform the world of them.[1]

His uncle Justin had died at the age of seventy-seven, after reigning nine years. His accession had marked a sort of resurrection in eastern affairs. Instead of three emperors, Basiliscus, Zeno, and Anastasius, alike ignominious in their government, unsound in their faith, infamous in their life, and remorselessly tyrannical in their treatment both of Church and State, Justin had crowned an honourable life as a general in the imperial service with a creditable reign, in which his fidelity to the Catholic faith was remarkable. The moment of Justinian's succession was coeval with great

---

[1] This paragraph translated from Rump, ix. 70.

changes in the West. By the death of Theodorick, who in his last year had begun the work of active Arian persecution, the great kingdom which he had maintained for a generation seemed on the point of dissolution, through the intrinsic inaptitude for government which his Gothic subjects at once betrayed when let loose from the master's powerful hand. In Africa, moreover, a succession of cruel Vandal persecutors, almost equal to their original, Genseric, had shaken their tenure of the country. At the same time, the Frankish kingdom, strengthened greatly by the conversion of Clovis, was growing in power and extent— a growth not interrupted by his early death in 511, at the age of forty-five.[1]

Such was the state of things when Justinian directed the great power which the revenues of the eastern empire enabled him to wield, towards the restoration of that empire, first in Africa, and then in Italy. Later in the same year, 533, in which he addressed to John II. the explicit acknowledgment of his supreme authority with which I began, he despatched his great general Belisarius with 16,000 chosen troops, 6000 of them cavalry, to Carthage. The Vandal ruler Gelimer offered but a feeble and utterly ineffectual resistance. He surrendered himself at Carthage to Belisarius, by the end of the year, and was brought to Constantinople. There Justinian received Belisarius in what was like one of Rome's hundred triumphs, except that the conqueror marched on foot. The booty of the Vandal kings was

[1] Rump, viii. 487.

borne before him, in which were conspicuous the precious things which Genseric had carried away from Rome—the vessels of the temple of Jerusalem. When the captive king was brought into the circus, and saw before him the emperor and countless rows of spectators, he is said to have shed no tears, but to have uttered the words of the preacher: "Vanity of vanities, all is vanity". But his head did not fall under the axe of the lictors, as in the ancient Roman triumphs. He received in Dalmatia a great property, and lived there in abundance with his family. The other captives were enrolled in the Roman army, and Justinian and Theodora heaped presents upon the daughters of Hilderich, and all the descendants of that princess Eudocia, great-granddaughter of the great Theodosius, who had been obliged to espouse the son of Genseric in her captivity at Carthage.

Then Justinian divided North Africa into seven provinces—Tingitana, Mauritanea, Numidia, Carthage, Byzacene, Tripolis, and Sardinia, which last, having belonged to the Vandals, was put into the prefecture of Africa. This received a Prætorian prefect and proconsular governors, who were charged to maintain the land, and show to the inhabitants the difference between civilised Roman government and Vandal cruelty. Justinian restored many cities, and erected many great buildings, especially churches, of which five in Leptis alone.[1]

An early result of Justinian's reconquest of Africa

---

[1] Account from Rump, ix. 172-4, compressed.

was that the bishops met in plenary council, under the presidency of the primate of Carthage, Reparatus, successor of Boniface. After a hundred years of Vandal oppression, 217 bishops assembled in the Basilica of Faustus, at Carthage, named Justiniana in honour of the emperor—the church which Hunnerich had taken from the Catholics, in which many bodies of martyrs were buried. To their intercession the council ascribed their deliverance from persecution. After reading the Nicene decrees, they discussed the question whether Arian priests who had become Catholics should be received in their dignity or only to lay communion. All the members of the council inclined to the latter judgment. They, however, would come to no decision, but with one voice determined to consult Pope John II. They addressed a letter to him by the hands of two bishops and a deacon, in which they say: "We considered it agreeable to charity that no one should disclose our judgment until first the custom or determination of the Roman Church should be made known to us: honouring herein with due obedience the authority of your Blessedness, being such a Pontiff as the holy See of Peter deserved to have, worthy of veneration, full of affection, speaking the truth without falsehood, doing nothing with arrogance. Therefore the free charity of the whole brotherhood thought that your counsel should be asked. And we beg that your mind, the organ of the Holy Spirit,[1] may answer us kindly and truly."[2]

---

[1] Respondeat mens illa Sancto Spiritui serviens.   [2] Mansi, viii. 808.

When the African deputies reached Rome, Pope John II. was already dead. But his successor Agapetus answered the questions of the council, attaching also the ancient canons which decided thereupon, to the effect that at whatever age a person had been infected by the Arian pestilence, if he became afterwards a Catholic he should not retain any rank, but that converted Arian priests might receive support from the Church fund. Pope Agapetus wrote expressing his intense joy at the recovery of their country: "For, since the Church is everywhere one body, your sorrow was our affliction. And we acknowledge your most sincere charity in that, as became wise and learned men, you did not forget the Apostolic Principate; but, in order to resolve that question, sought approach to that See to which the power of the keys is given".[1]

This council also sent an embassy to Justinian, beseeching him to restore the possessions and rights of the Church in Africa which the Vandals had taken away—a request which the emperor granted in an edict to his Prætorian prefect Salomo. And Agapetus expressly restored to the primate of Carthage any rights as metropolitan which the enemy had taken away.[2]

Thus the terrible persecution inaugurated by Genseric when the Vandal host lay around the deathbed of St. Augustine at Hippo in 430 came to an end. In the interval, the African church had suffered every ex-

---

[1] Mansi, viii. 849.
[2] See Baronius, A.D. 535, sec. 40; Hefele, ii. 736-8; Rump, ix. 174-6; *Novell.* xxxix. *De Africana Ecclesia.*

tremity of barbarian cruelty from the Arian invaders. At the end, the primate of Carthage, at the head of all the bishops of the several provinces, is found referring to the Pope, a subject of the Arian Theodatus, for guidance in the treatment of Arian priests and bishops who submitted to the Church. The Pope, on his side, acknowledges all the rights of the primate of Carthage which existed before the invasion. As to civil rights of property, the Byzantine conqueror restores the possessions of the Church which had been taken away by the Vandals.

By the restoration of the African province to the Roman empire and the Catholic faith Justinian won great renown. His accession had been welcomed with joy by the Catholic people. Full of great designs, he aimed at the extension of his realm, and endeavoured to advance the Christian cause by missions to countries as yet without the faith. Greatness and majesty are shown in all his creations.[1] In the year following the African reconquest Pope Agapetus wrote to him, praising his solicitude in maintaining the unity of the Church, and identifying the advance of his empire with the increase of religion.[2] The Pope adds that the emperor desired the profession of faith which he had sent to his predecessor Pope John II., and which had been confirmed by him, to be confirmed also by himself,

---

[1] Photius, i. 153-4: words of Hergenröther, who quotes eastern historians, who call him μεγαλοπρεπίστερος ἀνάκτων τῶν προτέρων . . . μεγαλουργὸς κράτωρ.

[2] Mansi, viii. 846.

for which " we praise you : we assent, not because we admit in laymen an authority to preach, but because, since the zeal of your faith is in accordance with the rules of our fathers, we confirm and give it force".

It is to be remembered that Pope Agapetus, elected in 535, was the subject of the Gothic king Theodatus, and as such was sent by him, under threats of death, in the winter of this year, on an embassy to Justinian. The purpose of Theodatus was to support his tottering throne by the intercession of the Pope. He had murdered at the lake of Bolsena the daughter and heiress of Theodorick, Amalasunta, who had made him king upon the untimely death of her son Athalarick in 534. He was secretly proposing to cede the Gothic kingdom of Italy to Justinian for a pension of 1200 pounds of gold. Thus Agapetus was sent to Constantinople in the winter of 535, as Pope John I. had been sent by Theodorick ten years before. He entered that city on the 20th February, 536; he died on the 22nd April following. In these two months the Pope, the subject of Theodatus, did great things. A certain Anthimus, a secret friend of the Monophysite heresy, had been brought, by the favour of the like-minded empress Theodora, from the see of Trebisond and put into that of Constantinople, having been able to impose himself upon the emperor as orthodox. Agapetus was received with the greatest honour, being only the second Pope who had visited Byzantium. He could not negotiate a peace for Theodatus; but archimandrites, priests, and monks besought him to proceed against

Anthimus as an interloper and teacher of error. Agapetus refused his communion to the new patriarch, required of him a written confession of faith, and return to his bishopric, which he had deserted contrary to the canons. The emperor, believing in the orthodoxy of his patriarch, took part at first against the Pope, and strove to overcome him both with threats and with presents. But Justinian, undeceived as to the orthodoxy of Anthimus, gave him up, and Pope Agapetus pronounced judgment of deposition upon him, and on the 13th March, 536, consecrated Mennas, who had been duly elected, to be bishop of Constantinople. He first required of him a written confession " to carry to Rome, to St. Peter".[1]

Soon after this the Pope died suddenly. The whole population at Constantinople attended his funeral. Never, it was said, had the mourning for a bishop or an emperor drawn together such a concourse of people. His body was carried back to Rome in triumph and buried in St. Peter's.

Pope Agapetus was succeeded in 536 by Pope Silverius, chosen under the influence of the Gothic king Theodatus. He was the last Pope so chosen; and the moment of his election is coincident with events destined to change permanently the material condition both of Rome and Italy.

Justinian had accomplished, with singular ease and rapidity, the first half of his design. This was the reunion of North Africa to his empire, and the restora-

---

[1] Photius, i. 160-2; Rump, ix. 181.

tion in it of the Catholic faith. The second part of his design was to accomplish the same double result for Rome and for Italy. He sent Belisarius, after the victory at Carthage, into Sicily, where Syracuse and Palermo were taken; and in the summer of 536 the great commander entered Italy, captured Naples, and advanced towards Rome on the Appian Road. So the Gothic war began. Theodatus was in Rome. The Gothic army in the Pontine marshes became aware of his incompetence and his secret treating with Justinian, deposed him, and elected Vitiges to be their king in his stead, by whose orders the fugitive was slain in his flight on the Flaminian Road. But Vitiges hastened to Ravenna, where he espoused the unwilling Matasunta, daughter of Amalasuntha, granddaughter of Theodorick. Four thousand Goths alone remained to cover Rome. Belisarius appeared before it. A deputation, supported by Pope Silverius, brought him the keys of the city. The garrison was too weak to defend it, and on the 9th December, 536, Belisarius took possession of Rome, at the head of the imperial troops, who had nothing Roman in them except the name. It was sixty years since Odoacer had caused the senate to declare a western emperor needless, and Rome, as to temporal rule, had fallen, first under the Herule, then under the Goth. The Romans welcomed Belisarius as a deliverer from the double yoke of the northern intruder and the Arian heretic.

For however Theodorick recognised, after the fury of the conflict with his brother-Teuton, the Herule

Odoacer, was over, the necessity of ruling with justice over Goth and Italian, however prosperous as to the maintenance of peace and internal order the great kingdom stretching from Illyricum to Southern Gaul had been, whatever support he had given to the maintenance of Roman law, custom, and institutions, there was not a Roman, from Symmachus and Boethius in the senate to the meanest inhabitant of Trastevere, who would not loathe the occupation of Rome and Italy by the Gothic invasion. The Goths were a people of remarkable courage and extraordinary force of body. But the feeling with which Italians and, above all, Romans would regard them as masters of their country and confiscators of its soil, can only be expressed by what the English would feel if a swarm of Zulus were to take possession of England. So, when Belisarius entered Rome, the Romans looked for their being replaced under the direct and lawful government of one who should be in deed and in truth a Roman prince, as Pope Felix had called the recreant Zeno, that is, the head of law, the supreme judge, the defender of the Church. This was what they looked for. I am about to mention what they found.

The empress Theodora had tried with all her wiles to set a Monophysite prelate on the Byzantine See.[1] Pope Agapetus had frustrated her plans by deposing Anthimus and consecrating Mennas in his place. But Theodora had not given up her intrigues, and she

---

[1] Photius, i. 163. The words which concern the conduct of Vigilius are taken from Cardinal Hergenröther. Baronius, A.D. 538, sec. 5, gives from Anastasius the words of the empress, and the Pope's answer, and the following narrative.

strove to involve in her net the Roman See itself. In the train of Agapetus at Constantinople was the ambitious deacon Vigilius. She sought to win him by promising him the Roman See. She offered him a great sum of money, and all her powerful support in attaining the papal dignity, if he would bind himself thereupon to abrogate the Council of Chalcedon, to enter into communion with Anthimus and Severus, and help them to recover the sees of Constantinople and Antioch. Vigilius agreed, and Theodora worked for the interests of her favourite by means of Antonina, wife of Belisarius. In the meantime, Silverius, as we have seen, had been chosen Pope in Rome, and Theodatus had exercised in his favour the influence which the Teuton rulers, whether styled Patricius or King, had claimed in the papal election since Odoacer. The empress invited the new Pope to come to Constantinople, or at least to restore her dear Anthimus. Silverius refused decidedly, though he was in the most dangerous position between the Greeks and the Ostrogoths, and even his personal liberty was in danger from Belisarius.

Pope Silverius continued to refuse submission to the wishes of the empress. The great commander sat in the Pincian palace in March, 537, scarcely three months after he had taken possession of Rome.[1] There he abased himself to carry out the commands of two shameless women, Theodora and Antonina. He caused Pope Silverius to be brought before him on a charge of writing treasonable letters to Vitiges. The Pope had

[1] Gregorovius, i. 372. See Liberatus, *Breviarium*, ch. xxii.

taken refuge at Santa Sabina on the Aventine. When brought before Belisarius, he found him sitting at the feet of Antonina, who reclined on a couch. The attending clergy had been left behind the first and second curtains. The Pope and the deacon Vigilius entered alone. "Lord Pope Silverius," said Antonina, "what have we done to thee and the Romans that thou wouldst deliver us into the hands of the Goths?" While she was heaping reproaches upon him, John, a sub-deacon of the first region, entered, took the pallium from his shoulders, and led him into another room, where he was stript of his episcopal vestments, the dress of a monk was put upon him, and his deposition was announced to the clergy. He was then banished to Patara in Lycia. All these intrigues had been unknown to Justinian. Afterwards, the bishop[1] of Patara went to him, and invoked before the emperor the judgment of God, saying there were many kings in this world, but not one set over the Church of the whole world, as was that bishop who had been expelled from his see. Justinian, hearing this, ordered Silverius to be taken back to Rome, and a true judgment of his case to be made. But then the Pope fell entirely into the hands of his rival Vigilius, who in the meantime had, by the help of Belisarius, got possession of the pontificate. Vigilius caused him to be deported to the island of Palmaria. There it is only known that he died in great misery, but with the crown of martyrdom.

This was the first act of that dominion, lasting more

---

[1] Liberatus, *Breviarium*.

than two hundred years, in which the Byzantine sovereigns were lords of Rome, as part of a reconquered province, and claimed to confirm the Papal elections, a claim set up by the Herule Odoacer, continued by Theodorick, inherited by Justinian.

When Belisarius occupied Rome he had only 5000 soldiers at his command. Vitiges, the new Gothic king, had gone to Ravenna, and made peace with the Franks by surrendering to them the southern provinces of France, held by Theodorick. He then levied the whole fighting force of the Goths, and, in March, 537, advanced from Umbria upon Rome at the head of 150,000 men. Belisarius, in the three months, had done his best to repair the walls, the towers, and the gates of the city. He had also laid up provisions. He dug trenches round the least defended spots, and had constructed great machines which shot bolts strong enough to nail an armoured man to a tree. Vitiges approached from the Anio, and made a desperate attempt to storm the city at once. Having failed in this, through the great courage and skill of Belisarius, and being unable, even with his vast host, to surround the city, he set up six fortified camps from the Flaminian Gate to that of Prœneste, and a seventh in the Neronian fields on the other side of the river, the plain which stretches from the Vatican to the Milvian bridge. The Goth cut off the fourteen aqueducts which supplied Rome with water. Those greatest monuments of imperial magnificence from that time have stretched their broken arches across the Campagna, the admiration and sorrow of every beholder in so many generations. What five hundred years of empire had done, the Goth, in his fury to recover the land which he had usurped, was able to ruin. The besiegers went on wasting the Campagna, and preventing the entrance of provisions into the city. Amid the increasing want, and the fear of worse, Vitiges in vain tried to seduce the Romans to revolt. Finding that Belisarius would not capitulate, he constructed great wooden towers, loftier than the walls, upon wheels, from which fifty men to each should direct battering-rams. Belisarius opposed him with like weapons. On the nineteenth day, the Goths poured out from their seven camps for a general storm. In a tremendous conflict, Belisarius beat back the invaders by counter sallies at the gates assailed. But at one point they all but succeeded. The Mausoleum of Hadrian formed part of the defence. Procopius, the eye-witness of this famous siege, and its narrator, says of it: "The tomb of the Roman emperor Hadrian lies outside the Aurelian Gate, a stone's-throw from the walls—a work of marvellous splendour. For it consists of huge blocks of Parian marble, fastened to each other without jointing from inside. It has four equal sides, each of them in length a stone's-cast. Its height exceeds that of the city walls. Upon it stand wonderful statues of men and horses." This is all that Procopius says. Up to this moment, full four centuries after the death of Hadrian, all the glories of Grecian art, which that imperial traveller over the world, from Newcastle to the cataracts of the Nile, could collect, had shone through the Roman sky on the monument, splendid as a palace and strong as a castle. On this fatal day of Rome's direst need they were hurled down upon the advancing Goth, whom the narrow streets had enabled to approach with scaling ladders. Statues of emperors, gods, and heroes hailed upon the northern giants; the works of Polycletus and Praxiteles were used for common stones upon invaders who despised art as well as letters; and a thousand years afterwards, when the building was finally formed into a castle, in digging the trenches the fragments of the Sleeping Faun were found, which had crushed some inglorious barbarian and saved Rome from capture.

But the storming, repulsed at every gate, cost Vitiges the flower of his host. Thirty thousand are said to have fallen, that being the number which Procopius records as derived from Gothic officers themselves; and greater, he says, was the number of wounded, when the deadly bolts from the machines of Belisarius mowed down their encumbered masses in flight.

The result of this great conflict was to weaken the Goths, to encourage the Romans, to make Belisarius confident of success. The siege lasted after this nearly a year. The extremity of hunger and misery was endured in the city. The supply of water was reduced to the cisterns and springs and the river. Vitiges at length occupied Porto, and cut off Rome from the sea. But the Goths also suffered terribly both from famine and from summer heat. The end of all was that, after a siege of a year and nine days, in which the Goths had fought 69 battles, Vitiges, in March, 538, drew off his diminished troops. One morning, Belisarius, from his Pincian palace, saw one-half of the remaining Goths on the other side of the Milvian bridge, and he forthwith ordered a sally upon their rear-guard. Vitiges left perhaps the half of his great host mouldering in the wasted, pestilent, deserted Campagna. He left also a city impoverished in numbers, full of sickness and misery. He had destroyed all the villas and dwellings of the Campagna; the churches of the Martyrs lay in heaps of ruins: from the Porta Salara to the Porta Nomentana hardly one stone upon another seems to have remained. Also Vitiges had ordered the senators whom he had left at Ravenna to be put to death. Only, during this siege, the basilicas of Rome's patron saints, which lay outside the walls, received no damage and were respected by the Goths.[1]

After this the storm of war drew off to the North. It continued with changing fortune in the provinces of Tuscany, Æmilia, the plain of the Po, the coasts of the Hadriatic. On the one side Franks and Burgundians took part; on the other side the soldiers of Belisarius were made up of all races from the East: not without skill in fight, but without discipline, under rival and quarrelling commanders. They pressed grievously on the land which they were sent to deliver. But the Goths grew weaker: they never recovered their losses before Rome. At last Belisarius got hold of Ravenna —not by capture, but after long negotiations, on both

[1] Reumont, ii. 49.

sides deceptive. Belisarius made the Goths believe that he would set himself at their head, and construct a new western empire. Vitiges, whether he trusted him or not, came to terms with him. Belisarius proclaimed Justinian emperor. The German realm seemed broken to pieces: only Verona, Pavia, and a portion of Liguria held out. A small part only of the army still carried the national banner. Then the conqueror, in 539, was recalled to Byzantium, to conduct the war against Persia. He left Italy almost subdued, and carried with him the captive king of the Goths, Vitiges, as in former years he had carried Gelimer, the captive king of the Vandals. This was in 539, thirteen years after Theodorick's death.

The first act of that fearful drama, the Gothic war, was over. But as soon as Belisarius disappeared, the Goths began to recover themselves. The generals of Justinian lived on plunder. In Totila arose a new Gothic leader, the bravest of the brave. At the end of the year 541 he marched out of Verona with only five thousand men, defeated the incapable and disunited Grecian captains, took city after city, passed the Apennines, passed near Rome, without assailing it. In this career of victory the Gothic king once approached that Campanian hill on which the great benefactor of the West, St. Benedict, was laying the foundations of the cœnobitic life. In the first instance, Totila tried to deceive the Saint. He dressed up a high officer as king, and sent him, with three of his chief counts in attendance, to personate himself. When

Benedict saw the Gothic train approaching he was seated, and as soon as they were within earshot, he cried out to the warrior pretending to be king: "Son, lay aside that dress which is not thine". The Goth fell to the ground in dismay, and returned to report his discomfiture to Totila, who then came himself. But when he saw Benedict seated at a distance he prostrated himself, and though Benedict thrice bade him arise, he continued prostrate. The Saint then came to him, raised him up, upbraided him with the acts which he had committed, and revealed to him the future concerning himself: "Many evils thou doest; many hast thou done. Put a curb at length on thine iniquity. Rome, indeed, thou shalt enter; the sea thou shalt pass. Nine years thou shalt reign; in the tenth thou shalt die."[1] The king was awe-struck. The savage in him was quelled by the speaker's sanctity. From this time forth he altered his conduct, and became more humane. In the capture of Naples shortly afterwards he showed by his merciful treatment the effect which the presence of St. Benedict had produced on him, as well as in the following years of his life. This interview took place in the year 542.

But Totila[2] so advanced in power that, in spite of Byzantine intrigue and jealousy, Belisarius, having happily concluded the Persian war, was sent back to the supreme command in Italy. He landed in Ravenna, but without army, war-material, or money. In the

---

[1] St. Gregory, *Dialogues*, ii. 14, 18.
[2] The following drawn from Reumont's narrative, ii. 50-6.

summer of 545, Totila, having subdued the land all about Rome, laid siege to Rome itself. Belisarius occupied Porto, and Totila set up his camp eight miles from Rome, commanding the Tiber, and turning the siege into the closest blockade. In vain Belisarius attempted to burst the Gothic bar of the river and introduce provisions to Rome. In vain embassies were sent to Constantinople for help. The most frightful distress ensued at Rome. At length, after about eighteen months, certain Isaurian soldiers of the Greek garrison gave up the Porta Asinaria, and on the night of the 17th December, 546, Totila took the ill-defended city. When he entered, it was almost without inhabitants. Those whom the sword, famine, and pestilence had not yet taken were in flight or hiding. Patricians crept about in the garb of slaves. The number of victims at this capture was small. The desolation and misery seem to have worked not only on Totila, but also on his army. The plunder, which a captured city could not escape, was generally bloodless; but many houses were burnt in the Trasteverine quarter. As Theodorick had offered his prayers at the tomb of the Apostles, so Totila went from the Lateran to St. Peter's. What a change had the forty-six years brought about. To the miserable remnant of the senate Totila upbraided the ingratitude which had been shown for Gothic benefits under Theodorick. He accepted, however, the intercession of the deacon Pelagius, and protected not only the female sex in general, but especially the noble Rusticiana, widow of Boethius and daughter of Symmachus. Amalasunta had restored their property to her sons, the younger Boethius and Symmachus; but the war seems to have consumed everything. She was now a beggar, and the wild host of Totila wished to put her to death for having, as she was charged, maimed statues of Theodorick. But the king rescued her from their fury.

In the first impulse of wrath Totila had threatened to level Rome with the ground. Belisarius, lying sick at Porto, had addressed to him a letter, entreating him to spare the greatest and noblest of cities. He did, however, throw down a considerable part of the walls, and when he marched to Lucania against the Greeks, took with him the chief citizens, and made the rest of the inhabitants migrate to Campania. He left a desert behind him. If we could trust the exaggerated reports of Greek historians, Rome remained forty days without inhabitants, tenanted only by beasts.

So ended the second act of the Gothic tragedy.

But as Vitiges had quitted Rome, so Totila deserted it, and in the spring of 547 it was entered again by Belisarius. In less than a month he restored as well as he could the part of the walls demolished, called back the inhabitants lingering in the neighbourhood, and prepared for a new attack. It was not long in coming. Scarcely had the gaps in the walls been filled up by stones piled in disorder and the trenches cleared, when the Gothic king reappeared. Thrice was his assault repulsed; then he gave up the attempt, broke down the bridges over the Anio behind him, and went to

Tibur, which he took by treachery of the inhabitants, who were at strife with the Isaurian garrison. Totila massacred the citizens, the bishop, and the clergy; got possession of the upper course of the Tiber, and cut off the Romans from Tuscany. But then Belisarius was enabled to give greater care to repairing the city's defences. The state in which several gates remain to this day still show his hand. He restored Trajan's aqueduct, which fed the mills on the right bank. But in the winter of 547 the great captain was drawn away from Rome to carry on a miserable petty war with insufficient force in the south of Italy, and was finally recalled to Constantinople. So ended the third act of Rome's fall.

But Totila hastened from place to place, from victory to victory. After scouring the South and then Umbria at the beginning of 549, he stood the third time before Rome. A strong Byzantine garrison in the city had provided magazines, and the wide spaces within the walls had been sown with wheat. His first attack failed; but treachery opened to him the Ostian gate, and its famished defenders soon surrendered the mausoleum of Hadrian. The conqueror, in this fourth capture of the city, acted mildly. He called back the yet absent inhabitants, amongst them many of the senators who had been sent into Campania. How had the nobles of Rome melted away! Vitiges had ordered those kept in Ravenna as hostages to be slain. Some had then escaped to Liguria. The distrust of the Greeks as well as of the Goths threatened them.

Cethegus, chief of the senate, had been compelled to leave before the first siege of Totila. Now Totila did not succeed in coming to terms with Justinian. The Greek army received a new commander in the eunuch Narses, who had served before under Belisarius. In him skill, energy, court favour, and the command of considerable forces were united. Before the end of 549, Totila left Rome. Almost all Italy save Ravenna was in his hands. He dealt generously with the people, whilst the Byzantine officials, exhausting the land with their exactions, added to the sufferings of war.

And now we reach the fifth act of the drama in which Rome was humbled to the very dust. Totila, for more than two years and a half, carried on an unceasing struggle over land and sea—Sicily, Sardinia, Corsica, which he subdued, and beyond the Hadriatic, to the opposite coasts. Though generally victorious, he was more like the leader in an old Gothic raid than a king who ruled and defended a great realm. At last, in the spring of 552, Narses advanced from Ravenna with a great force to a decisive battle for Rome. Totila advanced from Rome into Tuscany to meet him. At Taginas, on the longest day, the conflict which decided the fate of the Gothic kingdom took place. All that summer day the battle lasted. The Gothic king, a true knight in royal armour, on a splendid steed, marshalled and led his host. When night had come his cavalry was overthrown, his footmen broken. The spear of a Gepid had wounded him mortally. He was taken from the field, died in the night, was hastily buried. But his grave was disclosed to the Greeks. They left him where he lay; only his blood-stained mantle and diadem set with precious stones were carried to Constantinople. Six thousand of his bravest warriors lay on the field of battle. Yet when the remains of the host collected themselves in Upper Italy they elected Teia in Pavia for head of the yet unconquered race.

But Narses, having captured the strong places in Middle Italy, advanced upon Rome. The Gothic garrison was too weak to defend the wide circuits of the walls. Parts were soon taken. Presently Hadrian's tomb, which Totila had surrounded with fresh walls, alone held out. But it soon fell, and hapless Rome was captured for the fifth time in the reign of Justinian. It was a day of doom for the still remaining noble families. Goths and Greeks alike turned against them. In Campania and in Sicily many distinguished Romans had waited for better times. Now not only the flying Goths cut down all who fell into their hands, but the barbarian troops in the army of Narses, at their entrance into Rome, followed the example. Then, again, three hundred youths of the noblest families, who had been kept as hostages at Pavia, were all executed by Teia. The western consulate ended in 534, Flavius Theodorus Paulinus being the last. It continued seven years longer in the East, where to Flavius Basilius, consul in 541, no successor was given. When Justinian abolished this dignity it had lasted 1050 years, with few interruptions. Though for more than half this time it had been a mere title of honour, yet the consuls gave their name to the year, and served still, it may be, to mark to the world the unity of the Roman empire.

From Rome the conqueror Narses turned his steps southwards to Cumæ, that he might seize the treasure of the Goths, which was guarded by the new king Teia's brother Aligern. This brought Teia himself by a rapid march down the Hadriatic coast, and crossing Italy obliquely, he appeared at the foot of Vesuvius. There, in the spring of 553, Teia fought a last and desperate battle over the grave of sunken cities, in view of the Gulf of Naples. At the head of a small host, he fought from early morn to noon. It was like a battle of Homeric warriors. Then he could no longer support the weight of twelve lances in his shield, and, calling to his armour-bearer for a fresh shield, he fell transfixed by a lance. The next day the remnant of the army, save a thousand who fought their way through and reached Pavia, accepted terms from Narses, to leave Italy and fight no more against the emperor.

But Italy was far yet from tranquillity. Teia had incited the Alemans and the Franks to break into Italy. The two brothers, Leuthar and Bucelin, led a raid of 70,000 men, who ravaged Central and Southern Italy down to the Straits of Sicily. One of these barbarians carried back his spoil-laden troops to the Po, where pestilence consumed him and his horde. The host of the other brother, Bucelin, when it had reached Capua, was overthrown on the Vulturnus by Narses, with a slaughter as utter as that which Marius inflicted on the

Cimbri. Scarcely five are said to have escaped. So, in the spring of 555, after twenty years of destruction, ended the Gothic war.[1]

The reconquest of North Africa from the Vandals cost Justinian a few months of uninterrupted victory. The reconquest of Italy from the Goths cost twenty years of suffering to both sides, leaving, indeed, Justinian master but of a ruined Italy, master also of Rome, but after five successive captures; its senate reduced to a shadow, its patricians all but destroyed, its population shrunk, it is supposed, when Narses took possession of it in 552, to between thirty and forty thousand impoverished inhabitants. But the greatest change remains to be recorded. The Pope had indeed been delivered from Arian sovereigns, who held the country under military occupation, but exercised their civil rule with leniency and consideration, bearing, no doubt, in mind that they were, at least in theory, vice-gerents of an over-lord who ruled at Constantinople what was still the greatest empire of the world. What Pope Gelasius truly called "hostile domination" had been tempered during three-and-thirty years by the personal qualities of one who was at once powerful in arms and wise in statesmanship. Rome, in the time of Theodorick and Athalarick, had been maintained, its senate respected, the Pope treated with deference. A stranger entering Rome in 535, at the beginning of the Gothic war, would still have seen the greatest and grandest city of the world, standing in general with its buildings unimpaired.

[1] The narrative drawn from Reumont, ii. 56-7; Gregorovius, i. 448-9.

In 552, the Pope, instead of a distant over-lord, to whom he could appeal as Roman prince, had received an immediate master, who ruled Rome by a governor with a permanent garrison, and who understood his rule at Rome to be the same as his rule at Byzantium. The same as to its absolute power; but with this difference, that while Byzantium was the seat of his imperial dignity, in which every interest touched his personal credit, and its bishop was to be supported as the chief officer of his court and the chief councillor of his administration, the Rome he took from the Goths was simply a provincial town of a recovered province, once indeed illustrious, but now ruined and very troublesome. A provincial town because the seat of Byzantine power in Italy was henceforth not at Rome but at Ravenna, while the sovereign of Italy no longer held his court within Italy, at Ravenna or at Verona, as Theodorick and Athalarick, but at Constantinople. Mature reflection upon the civil condition made for the Pope by the result of the Gothic war will, I think, show that no severer test of the foundation of his spiritual authority could be applied than what this great event brought in its train. Nor must we omit to note that this test was brought about not only by the operation of political causes, but by actors who had not the intention of producing such a result. The suffering of Rome, in particular, during this war at the hands of Vitiges, Belisarius, Totila, Teia, Narses, is indescribable. It is hard to say whether defender or assailant did it most injury; but it is true to say that the one and the other were equally merciless in their purpose to retain it as a prey or to recover it as a conquest. Vitiges, besides pressing the people cooped up in its walls with a terrible famine during his siege of a year, broke down its aqueducts and ruined every building on that part of the Campagna which he scoured. Totila, in like manner, after famishing the inhabitants, when he took Rome, broke down a good part of its walls, and at his second capture, in 546, the city is described as having been absolutely deserted. In the last struggle, Teia slew without pity the three hundred hostages of Rome's noblest blood who had been sent to Pavia, thereby almost destroying its patricians. These were the parting tokens of Gothic affection for Italy. Then Belisarius, attempting to relieve Rome with inadequate forces, which was all that the penury of Justinian allowed him, was the means of prolonging the famine, while he did not save the city from capture. Lastly, Narses, sent to finish the war, enrolled in Dalmatia an army of adventurers. Huns, Lombards, Herules, Gepids, Greeks, and even Persians, in figure, language, arms, and customs utterly dissimilar, fought for him under the imperial standard, greedy for the treasures of Italy. Narses took Rome in 552, and governed it as imperial prefect for fifteen years at the head of a Greek garrison, until he was recalled in 567. That occupation of Narses in 552 is the date of Rome's extinction as the old secular imperial city. The year after his recal came the worst plague of all, and the most enduring. The Lombards did but repeat for the subjection of Italy to a fresh northern invasion what Narses had done to deliver it from Theodorick's older one in the preceding century.

Now let us see the nature of the test which this course of events, the work of Goth and Greek alike—inflicting great misery and danger on the clergy and the Pope, as upon their people—applied to the papal authority itself.

A more emphatic attestation of that authority than the confession given in 519 to Pope Hormisdas by the whole Greek episcopate, and by the emperor at the head of his court, could hardly be drawn up. It settled for ever the question of right, and estopped Byzantium, whether in the person of Cæsar or of patriarch, from denial of the Pope's universal pastorship, as derived from St. Peter. We have seen that not only did Justinian, when the leading spirit in his uncle's freshly-acquired succession to the eastern empire, do his utmost to bring about this confession, but that in the first years of his reign his letter to Pope John II. reaffirmed it; and his treatment of Pope Agapetus when he appeared at Constantinople, not only as Pope, but in the character of ambassador from the Gothic king Theodatus, exhibited that belief in action. But now a state of things quite unknown before had ensued. Hitherto Rome had been the capital, of which even Constantine's Nova Roma was but the pale imitation. But the five times captured, desolate, impoverished Rome which came back under Narses to Justinian's sway, came back not as a capital, but as a captive governed by an exarch. Was the bishop of a city with its senate extinct, its patriciate destroyed, and with forty thousand returned refugees for its inhabitants, still the bearer of Peter's keys — still the Rock on which the City of God rested? Had there been one particle of truth in that 28th canon which a certain party attempted to pass at the Council of Chalcedon, and which St. Leo peremptorily annulled, a negative answer to this must now have followed. That canon asserted "that the Fathers justly gave its prerogatives to the see of the elder Rome because that was the imperial city". Rome had ceased to be the imperial city. Did the loss of its bishop's prerogatives follow? Did they pass to Byzantium because it was become the imperial city, because the sole emperor dwelt there? Thus, about a hundred years after the repulse of the ambitious exaltation sought by Anatolius, its rejection by the provident wisdom and resolute courage of St. Leo was more than justified by the course of events. St. Leo's action was based upon the constitution of the Church, and therefore did not need to be justified by events. But the Divine Providence superadded this justification, and that under circumstances which had had no parallel in the preceding five hundred years.

For when Belisarius, submitting himself to carry out the orders of an imperious mistress, deposed, as we have seen, the legitimate Pope Silverius by force in March, 537, Vigilius, in virtue of the same force, was consecrated a few days after to succeed him. The exact time of the death which Pope Silverius suffered in Palmeria is not known. But Vigilius is not recognised as lawful Pope until after his death, probably in 540. He then ascended St. Peter's seat with a blot upon him such as no pontiff had suffered before. And this pontificate lasted about fifteen years, and was full of such humiliation as St. Peter had never suffered before in his successors.

We are not acquainted with the detail of events at Rome in those terrible years, but we learn that, as Pope John I. was sent to Constantinople as a subject by Theodorick, and Pope Agapetus again as a subject by Theodatus, so Vigilius was urged by Justinian to go thither, and that after many delays he obeyed the emperor very unwillingly.

But it is requisite here to give a short summary of what Justinian had been doing in the affairs of the eastern Church from the time that Pope Agapetus, having consecrated Mennas to be bishop of Constantinople, died there in 536. After the Pope's death, Mennas proceeded to hold in May and June of that year a synod in which he declared Anthimus to be entirely deposed from the episcopal dignity, and condemned Severus and other leaders of the Monophysites. In this synod Mennas presided, and the two Roman deacons, Vigilius and Pelagius, who had been the legates of Pope Agapetus, but whose powers had expired at his death, sat next to him, but only as Italian bishops. How little the patriarch Mennas could there represent the Church's independence is shown by his words to the bishops in the fourth session: "Your charity knows that nothing of what is mooted in the Church should take place contrary to the decision and order of our emperor, zealous for the faith," while of their relation to the Pope he said: "You know that we follow and obey the Apostolic See; those who are in communion with it we hold in communion; those whom it condemns we also condemn".[1] Justinian, irritated by the boldness of the Monophysites, added the sanction of law to the decrees of this council, which deposed men who had occupied patriarchal sees. He used these words: "In the present law we are doing an act not unusual to the empire. For as often as an episcopal decree has deposed from their sacerdotal seats those unworthy of the priesthood, such as Nestorius, Eutyches, Arius, Macedonius, and Eunomius, and others in wickedness not inferior to them, so often the empire has agreed with the authority of the bishops. Thus the divine and the human concurred in one righteous judgment, as we know was done in the case of Anthimus of late, who was deposed from the see of this imperial city by Agapetus, of holy and renowned memory, bishop of Old Rome."[2]

In the intrigue of Theodora with Vigilius, Mennas took no part. He took counsel with the emperor how to maintain the Catholic faith in Alexandria against the heretical patriarch Theodosius. By the emperor's direction, ordering him to expel Theodosius, Mennas, in 537 or 538, consecrated Paul, a monk of Tabenna, to be

---

[1] Mansi, viii. 969; Photius, i. 163.   [2] Mansi, viii. 1149.

patriarch of Alexandria. The act would appear to have been done in the presence of Pelagius, then nuncio in Constantinople, without reclamation on his part, or of the nuncios who represented Antioch and Jerusalem. Mennas in this repeated the conduct of Anatolius and Acacius in former times, who were censured, the one by St. Leo, the other by Pope Simplicius. By this event the four eastern patriarchs seemed to agree to accept the first four councils, and the unity of the Church to be quite restored, from which Alexandria had until then stood aloof; but the patriarch Paul came afterwards in suspicion of heresy and had to give way to Zoilus. Mennas was on the best terms with the emperor; he might easily have used the deposition of Silverius and the unlawful exaltation of Vigilius in 537 for increase of his own influence, had not a feeling of duty or love of peace held him back. But Vigilius also, when he came to be acknowledged, had come to realise his position and its responsibility. He was far from fulfilling the unlawful promises made to Theodora, and from favouring the Monophysites. The empress found that she had thrown away her money and failed in her intrigue. In letters[1] to the emperor and to Mennas, in 540, Vigilius declared his close adherence to the acts of his predecessors, St. Leo in particular, and to the decrees in faith of the four General Councils, while he confirmed the acts of the council held by Mennas against Severus and the other Monophysite leaders.

In the meantime new dissensions threatened to

[1] Mansi, ix. 35-40.

agitate the whole eastern realm.[1] The partisans of Origen in Palestine and the neighbouring countries rose. At their head stood Theodore Askidas, archbishop of Cæsarea in Cappadocia, and Domitian, metropolitan of Ancyra, who had obtained, by favour of Justinian, these important sees. Ephrem, patriarch of Antioch about 540, condemned Origenism in a synod. Pelagius, being papal nuncio at Constantinople, had, together with Ephrem, patriarch of Antioch, condemned the patriarch Paul of Alexandria at Gaza. Deputies from Peter, patriarch of Jerusalem, and the orthodox monks journeyed with Pelagius to Constantinople, to present to the emperor an accusation against the Origenists. Pelagius had much influence with Justinian, and he and Mennas procured for the petitioners access to the emperor. They asked him to issue a solemn condemnation of Origen's errors. The emperor listened willingly, and issued in the form of a treatise to Mennas a still extant censure of Origen and his writings. He called upon the patriarchs to hold synods upon them. Mennas, in 543, held one in the capital, which issued fifteen anathemas against Origen.[2] Theodore Askidas and Domitian, by submitting to the imperial edict and the condemnation of Origen, kept their places and secured afresh their influence, which the monks of Palestine, who were not Origenistic, felt severely. They even managed, in the interest of their party, to turn the attention of the dogmatising emperor to

---

[1] Narrative drawn from Photius, i. 165-6, down to "Ferrandus," p. 232, below.
[2] Mansi, ix. 487-537.

another question, and moved him to issue, in 544, the edict upon the Three Chapters. He thought he was bringing back the Monophysites to orthodoxy. He was really casting a new ferment into the existing agitation.

At first the patriarch Mennas was very displeased with this edict censuring in the so-called Three Chapters Theodoret, Ibas, and Theodore of Mopsuestia as Nestorians. He considered the credit of the Council of Chalcedon to be therein impeached, and declared that he would only subscribe to it after the Pope had subscribed. Afterwards, being more strongly pressed, he subscribed unwillingly, but with the reservation, confirmed to him even upon oath, that if the Bishop of Rome refused his assent his signature should be returned to him, and his subscription be regarded as withdrawn. The other eastern patriarchs also at first resisted, but finished by complying with the imperial threats, as particularly Ephrem of Antioch. Most of the bishops, accustomed to slavish subjection to their patriarchs, followed their example, and Mennas had to urge the bishops under him by every means to comply. However, many bishops complained of this pressure to the papal legate Stephen, who pronounced against the edict, which seemed indirectly to impeach the authority of the Fourth Council. He even refused communion with Mennas because he had broken his first promise and given his assent before the Pope had decided upon it. Through the whole West the writings of Theodore, Theodoret, and Ibas were little known, but the decrees of Chalcedon were zealously maintained. The edict was refused, especially in Northern Africa. It was censured by the bishop Portian in a writing addressed to the emperor, and by the learned deacon Ferrandus.

Means had been taken by fraud and force to win the whole East to consent to the edict.[1] Mennas, patriarch of Constantinople; Ephrem, patriarch of Antioch; Peter, patriarch of Jerusalem, crouched before the tyranny of Justinian; and so also Zoilus of Alexandria, though he promised Vigilius that he would not sign the edict, afterwards subscribed it.[2] At this point Justinian sought before everything to get the assent of the Pope, and he sent for Vigilius to Constantinople. He claimed the presence of Vigilius as his subject in virtue of the conquest of Belisarius: he meant to use this authority of Vigilius as Pope for his own purpose. Vigilius foresaw the difficulties into which he would fall. At length he left Rome in 544, before Totila began the second siege. He lingered in Sicily a year, in 546; he then travelled through Greece and Illyricum. At last he entered Byzantium on the 25th January, 547, and was welcomed with the most brilliant reception. Justinian humbly besought his blessing, and embraced him with tears. But this good understanding did not last long. Vigilius approved the conduct of his legates and refused his communion to Mennas, who, in signing the formula of Hormisdas, had bound himself to follow the Roman See, and had broken his special promise. Vigilius

---

[1] Hefele, ii. 790.
[2] Hergenröther, *K. G.*, i. 344-5; Photius, i. 166.

withdrew it also from the bishops who had subscribed the imperial edict. He and the bishops attending him saw in this edict a scheme to help the Acephali, upon whom Vigilius repeated his anathema. But Mennas feared the emperor much more than he feared the Pope, whose name he now removed from commemoration at the Mass. Vigilius, like the westerns in general, considered the edict to be useless and dangerous, as giving a pretext for seeming to abrogate the Council of Chalcedon, and also as a claim on the part of the emperor to the highest authority in Church matters. Justinian tried repeatedly his personal influence with the Pope, that also of bishops and officers of State. He even had him watched for a length of time and cut off from all approach, so that the Pope exclaimed, "If you have made me a prisoner, you cannot imprison the holy Apostle Peter". Yet the intercourse of Vigilius with eastern bishops soon convinced him that they were generally agreed with the emperor; that a prolonged resistance on his part would produce a new division between Greeks and Latins; that considerable grounds existed for the condemnation of the Three Chapters, with which, hitherto, he had not been well acquainted. So he allowed the subject to be further considered, held out a prospect of agreeing with the emperor, and re-admitted Mennas to his communion, who restored the Pope's name in the liturgy. This reconciliation took place on the feast of the Princes of the Apostles, 29th June, 547.

The Pope, after further conferences with bishops

present at Constantinople, seventy of whom had not signed the imperial edict, issued, on the 11th April, 548, his *Judgment*, directed to Mennas, of which all but fragments are lost. In it he most strongly maintained the authority of the four General Councils, especially of the fourth; put under anathema the godless writings of Theodore of Mopsuestia, and also his person; the letter said to be written by Ibas to Maris, which Justinian had marked as supposititious, and the writings of Theodoret, which impugned orthodoxy and the twelve anathemas of Cyril. It was his purpose to quiet excitement, satisfying the Greeks by a specific condemnation of the Three Chapters, and the Latins by maintaining the rank of the Council of Chalcedon. And he required that therewith the strife should cease. But neither side accepted the condition. The westerns, especially Dacius, archbishop of Milan, and Facundus, bishop of Hermiane, vehemently attacked his *Judgment*. So did many African monks. Even two Roman deacons, the Pope's own nephew Rusticus, and Sebastianus, though they began by supporting the *Judgment*, became very violent against the Pope, spread the most injurious reports against him, and disregarded his warnings. He deposed and excommunicated them. False reports were spread that, against the Council of Chalcedon, the Pope had condemned the persons of Theodoret and Ibas, and had gone against the decrees of his predecessors. The Pope, after the death of the empress Theodora, on the 28th June, 548, had continued by the emperor's wish at Constantinople, especially since Totila had retaken Rome in 549. He had gone to Thessalonica and returned; he tried in several letters to the bishops of Scythia and Gaul to correct their misconceptions. These, however, prevailed with the bishops of Illyria, Dalmatia, and Africa, who in 549 and 550 separated themselves from the communion of Vigilius. A thing not heard of before now occurrred. The Roman Bishop stood with the Greek bishops on one side, the Latin bishops on the other, and the bewilderment increased from day to day.

In the summer of 550 the Pope and the emperor came to an agreement that a General Council should be held at which the western bishops should be present, until which all dispute about the Three Chapters, and any fresh step on the subject, should be forbidden, and in the meantime the Pope's *Judgment* should be returned to him. That took place at once, and preparations were made for the council. In June a council held at Mopsuestia by direction of the emperor declared that from the time of human memory the name of its former bishop, Theodore, had been erased from commemoration, and the name of St. Cyril put in. But the western bishops avoided answering the invitation to the council. The Illyrian did not come at all; the African sent as deputies Reparatus, the primate of Carthage, Firmus of Numidia, and two Byzacene bishops. These were besieged both with threats and presents; two were induced to sign the imperial edict; the other two were banished, Reparatus under charge of a political crime. While the western bishops showed still less inclination to appear, the court broke its agreement with Vigilius. A new writing against the Three Chapters was read in the palace before several bishops, and subscribed by them. Theodore Askidas, the chief contriver, and his companions, excused themselves to the Pope, who called them to account, and begged pardon, but spread the writing still more, set the emperor against Vigilius, and induced him to publish, in 551, a further edict under the name of a confession of faith. It contained, together with a detailed exposition of doctrine upon the Trinity and Incarnation, thirteen anathemas, with the refutation of different objections made by the defenders of the Three Chapters; for instance, that the letter of Ibas had been approved at Chalcedon, the condemnation of dead men forbidden, and Theodore of Mopsuestia been praised by orthodox Fathers.

The restoration of peace was thus made much more difficult, and the promise given to the Pope broken. The Pope protected himself against this violation of the agreement, by which nothing was to be done in the matter before the intended council, and considered himself released from his engagements. He saw herein the arbitrary interference of a despotic ruler anticipating the council's decision, which put in question the Church's whole right of authority, and much increased the danger of a schism. In an assembly of Greek and Latin bishops held in the Placidia palace, where he resided, he desired them to request the emperor to withdraw the proposed edict, and to wait for a general consideration of the subject, and especially for the sentence of the Latin bishops. If this was not granted, to refuse their subscription to the edict. Moreover, the See of Peter would excommunicate them. Dacius, also, archbishop of Milan, spoke in this sense. But the protest was disregarded, and Theodore Askidas, who had formed part of the assembly, went with the bishops of his party to the Church in which the edict was posted up, held solemn service there, struck out of the diptychs the patriarch Zoilus of Alexandria, who declined to condemn the Three Chapters, and proclaimed at once Apollinaris for his successor, with the consent of the weak Mennas, and in contempt of the Pope's authority. Not only now were the Three Chapters in question, but the whole right and independence of the Church's authority. Vigilius, having long warned the vain court-bishop Theodore Askidas, always a non-resident in his diocese, and having now been witness of a violence so unprecedented, put him under excommunication.

At this resistance Justinian was greatly embittered, and was inclined to imprison the Pope and his attendants. The Pope took refuge in the Church of St. Peter, by the palace of Hormisdas. He repeated with greater force his former declaration, entirely deprived Theodore Askidas, and put Mennas and his companions under ban, until they made satisfaction, on the 14th August, 551. At least the sentence was kept ready for publication. He was attended by eleven Italian and two African bishops. The emperor sent the prætor with soldiers to remove him by force. Vigilius clung to the altar, so that it was nearly pulled down with him. His imprisonment was prevented by the crowd which burst in, indignant at the ill-treatment offered to the Church's first bishop, and by the disgust of the soldiers at the gaol-work put upon them. The emperor, seeming to repent his hastiness, sent high officers of State to assure the Pope of personal security, at first with the threat to have him removed by force if he was not content with this; then he empowered the officers to swear that no ill should befal him. The Pope thereon returned to the palace of Placidia. But there, in spite of oaths, he was watched, deprived of his true servants, surrounded with paid spies, attacked with every sort of intrigue, even his handwriting forged. Then, seeing his palace entirely surrounded by suspicious persons, he risked, on the 23rd December, 551, a flight across the Bosphorus to the Church of St. Euphemia in Chalcedon, in which the Fourth Council had been held. Here, in January, 552, he published his decree against Theodore and Mennas, and was for a long time sick. When the emperor, with the offer of another oath, sent high officials to invite him to return to the capital, he replied that he needed no fresh oaths if the emperor had only the will to restore to the Church the peace which she enjoyed under his uncle Justin. He desired the emperor to avoid communion with those who lay under his ban. In his Encyclical of the 5th February, 552, he made known to all the Church what had passed, and expressed his belief and his wishes. Even in his humiliation the successor of Peter inspired a great veneration. They tried to approach him. He soon received a writing from Theodore Askidas, Mennas, Andrew, archbishop of Ephesus, and other bishops, in which they declared their adherence to the decrees of the four General Councils which had been made in agreement with the legates of the Apostolic See, as well as to the papal letters. They consented also to the withdrawal of all that had been written on the Three Chapters, and besought the Pope to pardon as well their intercourse with those who lay under his ban as the offences committed against him, in which also they claimed to have had no part. So things were brought to the condition in which they were before the appearance of the last imperial edict. Vigilius now returned from Chalcedon to Constantinople.

Mennas, who died in August, 552, was succeeded by Eutychius. He addressed himself to the Pope on the 6th January, 553, whose name had been restored by Mennas to the first place in the diptychs. Eutychius presented his confession of faith. He also proposed that a decision, in respect of the Three Chapters in accordance with the four General Councils, should be made in a meeting of bishops under the Pope's presidency. Apollinaris of Alexandria, Domnus of Antioch, Elias of Thessalonica, and other bishops subscribed this request. The Pope, in his reply of the 8th January, praised their zeal, and accepted the proposition of a council which he had before approved. Negotiations then began about its management. Here the emperor resisted the Pope's proposals in many points. He would not have the council held in Italy or Sicily, as the Pope desired, nor carry out his own proposal to summon such western bishops as the Pope named. He proposed further that an equal number of bishops should be consulted on both sides; hinting, moreover, that an equal number should be drawn from each patriarchate, while Vigilius meant an equal number from the East and the West, which he thought necessary to bring about a successful result. At last the emperor caused the council actually to meet on the 5th May, 553, under the presidency of Eutychius, with 151 bishops, among whom only six from Africa represented the West, against the Pope's will, in the secretarium of the chief church of Constantinople. First was read an imperial writing of much detail, which entered into the previous negotiations with Vigilius; then the correspondence between Eutychius and the Pope. It was resolved to invite him again. Vigilius refused to take part in the council, first on account of the excessive number of eastern bishops and the absence of most western; then of the disregard shown to his wishes. Further, he sought to preserve himself from compulsion, and maintain his decision in freedom. He had reason to fear the infringement of his dignity. Moreover, no one of his predecessors had taken personally a part in eastern councils, and Pope Celestine had forbidden his legates to enter into discussion with bishops, and appear as a party. The Pope maintained his refusal not only to the high officers of the emperor, but to an embassy from the council, at the head of which stood three eastern patriarchs. This he did, being the emperor's subject; being also in the power of an emperor who was able to appear to the eastern bishops almost the head of the Church, and to sway them as he pleased. The Pope would only declare himself ready to give his judgment apart. An account of this unsuccessful invitation was given in the council's second session of the 8th May. The western bishops still in the capital were invited to attend, but several declined, because the Pope took no part. At the third session, of the 9th May, after reading the former protocols, a confession of faith entirely agreeing with the imperial document communicated four days before was drawn up, and a special treatment of the Three Chapters ordered for another day. At the fourth session, seventy-one heretical or offensive propositions of Theodore of Mopsuestia were read and condemned. In the fifth, the opposition made to him by St. Cyril and others was considered, as well as the question whether it is allowable to anathematise after their death men who have died in the Church's communion. This was affirmed according to previous examples, and testimony from Augustine, Cyril, and others. Theodoret's writings against Cyril were also anathematised. In the sixth session, the same was done with the letter of Ibas. In the seventh session, several documents sent by the emperor were read, specially letters of Pope Vigilius up to 550, and a letter from the emperor Justin to his prefect Hypatius, in 520, forbidding that a feast to Theodore or to Theodoret should any longer be kept in the city of Cyrus. The imperial commissioner informed the council, likewise, that the Pope had sent by the sub-deacon Servusdei a letter to the emperor, which the emperor had not received, and therefore not communicated to the council. The longer Latin text of the acts also says that the emperor had commanded the Pope's name to be erased from the diptychs, without prejudice, however, to communion with the Apostolic See, which the council accepted. It held its last sitting on the 2nd June, 553, and issued fourteen anathemas in accordance with the thirteen of Justinian. There were then present 165 bishops.

The document brought to the emperor by the subdeacon in the Pope's name, but rejected, must be what has come down to us as the Constitution of the 14th May. It had the subscription of Vigilius, of sixteen bishops—nine Italian, three Asiatic, two Illyrian, and two African—with three Roman clergy. It decidedly rejected sixty propositions drawn from the writings of Theodore; anathematised five errors as to the Person of Christ; forbade the condemnation of Theodore's person, and of the two other Chapters. If this document was really drawn up by Vigilius, who had persisted during almost six years, as the emperor admitted, in condemning the Three Chapters, it must be explained by the Pope finding his especial difficulty in the manner of terminating the matter, so that the western bishops should be entirely satisfied that the decrees of the Council of Chalcedon remained inviolate; that he purposed only to condemn errors, but spare persons; that he wished to set his refusal against the pressure of the changeable emperor and the blind submission of the Grecian bishops, without surrendering any point of faith. Many irregularities appeared in what preceded the council and took place in it. Justinian's conduct was dishonouring to the Church, and he used force to get the decrees of the Council accepted. At last Vigilius, who seems with other bishops to have been banished, gave way to the pressure, and issued a decided condemnation of the Three Chapters, in a writing to Eutychius of 8th December, 553; and in a Constitution dated 23rd February, 554, he made no mention of the council, but gave his own decision in accordance with it, and independent of it, as he had before intended. Only by degrees the council held by Eutychius obtained the name of the Fifth General Council.

In August, 554, the Pope was again on good terms with the emperor, who issued at his request the Pragmatic Sanction for Italy. Then Vigilius set out to return to Rome, but died on his way at Syracuse in the beginning of 555. He had spent seven years in the Greek capital, in a position more difficult than had ever before occurred; ignorant himself of the language; struggling to his utmost to meet the dangers which assaulted the Church from every side. Now one and now another seemed to threaten the greater evil. He never wavered in the question of faith itself, but often as to what it was opportune to do: as whether it was advisable or necessary to condemn persons and writings which the Council of Chalcedon had spared: whether to issue a judgment which would be looked upon by the Monophysites as a triumph of their cause: which for the same reason would be utterly detested by most westerns, as a supposed surrender of the Council of Chalcedon; which, instead of closing the old divisions, might create new. Subsequent times showed the correctness of his solicitude.[1]

The patriarch Eutychius who presided at this council by the emperor's order, without the Pope, was held in great consideration by Justinian, and was consulted in his most important affairs. When Justinian had restored with the greatest splendour the still existing Church of Santa Sophia, Eutychius consecrated it in his presence on the 24th December, 563. Justinian then allotted to the service of the cathedral 60 priests, 100 deacons, 90 sub-deacons, 110 lectors, 120 singers, 100 ostiarii, and 40 deaconesses, a number which much increased between Justinian and Heraclius.

Justinian in his last years was minded to sanction by a formal decree a special doctrine which, after long resisting the Eutycheans, he had taken from them. It was that the Body of Christ was from the beginning incorruptible, and incapable of any change. He willed that all his bishops should set their hands to this decree. Eutychius was one of the first to resist. On the 22nd January, 565, he was taken by force from

[1] Translated from Hergenröther's *K. G.*, i., pp. 345-351, from p. 232, above, "at this point Justinian sought," &c., with reference also to the life of Photius.

his cathedral to a monastery; he refused to appear before a resident council called by the emperor, which deposed him, and appointed a successor. He was banished to Amasea, where he died, twelve years afterwards, in the monastery which he had formerly governed.[1]

But Justinian had become again, by the conquest of Narses, lord of Rome and Italy, and as such, in the year 554, issued at the request of Vigilius his Pragmatic Sanction. In Italy the struggle was at an end; the land was a desert. Flourishing cities had become heaps of smoking ruins. Milan had been destroyed. Three hundred thousand are said to have perished there. Before the recal of Belisarius, fifty thousand had died of hunger in the march of Ancona. Such facts give a notion of Rome's condition. In 554, Narses returned, and his victorious host entered, laden with booty, crowned with laurels. It was his task to maintain a regular government, which he did with the title of Patricius and Commander.[2] The Pragmatic Sanction was intended to establish a new political order of things in Italy, which was reunited to the empire. The two supreme officials of the Italian province were the Exarch and the Prefect. The title of Exarch then came up, and continued to the end of the Greek dominion in Italy. He united in himself the military and civil authority; but for the exercise of the latter the Prefect stood at his side as the first civil officer.

---

[1] Hergenröther, Photius, i. 174; Rump, *K. G.*, ix. 283.
[2] See Reumont, ii. 58-62; Gregorovius, i. 453-9.

Obedience to the whole body of legislation, as codified by Justinian's order, was enacted. For the rest the provisions of Constantine were followed. The administration of justice was in the hands of provincial judges, whom the bishops and the nobility chose from the ranks of the latter. It was then the bishops began to take part in the courts of justice of their own cities, as well in the choice and nomination of the officers as in their supervision.[1] The words Roman commonwealth, Roman emperor, Roman army, were heard again. But no word was said of restoring a western emperor. Rome retained only an ideal precedence; Constantinople was the seat of empire. Rome received a permanent garrison, and had to share with Ravenna, where the heads of the Italian government soon permanently resided. Justinian's constitution found existing the mere shadow of a senate. The prefect of the city governed at Rome. There is mention made of a salary given to professors of Grammar and Rhetoric,[2] to physicians and lawyers; but it is doubtful whether this ever came into effect. The Gothic war[3] seems to have destroyed the great public libraries of Rome, the Palatine and Ulpian, as well as the private libraries of princely palaces, such as Boethius and Symmachus possessed. And in all Italy the war of extermination between Goths and Greeks swallowed up the costly treasures of ancient literature, save such remnant as the Benedictine monasteries were able to collect and preserve.[4] No building of Justinian's in Rome is

---

[1] Reumont, 60.   [2] Gregorovius, 455.   [3] *Ibid.*, 456.   [4] Reumont, 61.

known. All his work of this kind was given to Ravenna. From this time forth every new building in Rome is due to the Popes.

Small reason had the Popes to rejoice that the rule of an orthodox emperor had followed at Rome that of an Arian king. Three months after the death of Vigilius at Syracuse Justinian caused the deacon Pelagius to be elected: he had difficulty in obtaining his recognition until he had cleared himself by oath in St. Peter's of an accusation that he had hastened his predecessor's death. The confirmation of the Pope's election remained with the emperor. This permanent fetter came upon the Popes from the interference of Odoacer the Herule in 484. After Justinian's death, the Romans sent an embassy to his successor complaining that their lot had been more endurable under the dominion of barbarians than under the Greeks.

When Narses,[1] re-entering Rome, celebrated a triple triumph over the expulsion of barbarians from Italy, the reunion of the empire, and the Church's victory over the Arians, a contempory historian writes that the mind of man had not power enough to conceive so many reverses of fortune, such destruction of cities, such a flight of men, such a murdering of peoples, much less to describe them in words. Italy was strewn with ruins and dead bodies from the Alps to Tarentum. Famine and pestilence, following on the steps of war, had reduced whole districts to desolation. Procopius compares the reckoning of losses to that of reckoning

[1] Gregorovius, 450-2.

the sands of the sea. A sober estimate computes that one-third of the population perished, and the ancient form of life in Rome and in all Italy was extinct for ever.

But before we make an estimate of Justinian's whole action and character and their result, a subject on which we have scarcely touched has to be carefully weighed.

What was the relation between the Two Powers conceived in the mind of Justinian, expressed in his legislation, carried out in his conduct, whether to the Roman Primate or the patriarchs of Alexandria, Antioch, Jerusalem, and Constantinople in his own eastern empire, or to the whole Church when assembled in council, as at Constantinople in 553? Was he merely carrying on as emperor a relation which he had inherited from so many predecessors, beginning with Constantine, or did he by his own laws and conduct alter an equilibrium before existing, and impair a definite and lawful union by transgressing the boundaries which made it the co-operation of Two Powers.

If we look back just a hundred years before his *Digest* appeared, we find, in the great deed[1] in which the emperors Theodosius II. and Valentinian III. convoked the Council of Ephesus, the charge which they considered to be laid upon the imperial power to maintain that union of the natural and the spiritual government on which, as on a joint foundation, the Roman State, in the judgment of its rulers, was itself built. Some of the words they use are: "We are the

[1] See vol. v. 281.

ministers of Providence for the advancement of the commonwealth, while, inasmuch as we represent the whole body of our subjects, we protect them at once in a right belief and in a civil polity corresponding with it".

This first and all-embracing principle of protecting all and every power which existed in the commonwealth, and maintaining it in due position, was most firmly held by Justinian. As to his own imperial authority and the basis on which it rested, he says: "Ever bearing in mind whatever regards the advantage and the honour of the commonwealth which God has entrusted to our hands, we seek to bring it to effect".[1] As to the Two Powers themselves, he recognises them thus: "The greatest gifts of God to men bestowed by the divine mercy are the priesthood and the empire; the former ministering in divine things, the latter presiding over human things, and exerting its diligence therein. Both, proceeding from one and the same principle, are the ornament of human life. Therefore nothing will be so great a care to emperors as the upright conduct of bishops, for, indeed, bishops are ever supplicating God for emperors. But if what concerns them be entirely blameless and full of confidence in God, and if the imperial power rightly and duly adorn the commonwealth entrusted to it, an admirable agreement will ensue, conferring on the human race all that is for its good. We then bear the greatest solicitude for the genuine divine doctrine, and for the upright conduct of bishops, which we trust, when that doctrine

[1] *Constitutio*, lxxxii. 667.

is maintained, because through it we shall obtain the greatest gifts from God,[1] shall be secure in the possession of those which we have, and shall acquire those which have not yet come. But all will be done well and fittingly if the beginning from which it springs be becoming and dear to God. And this we are confident will be, provided the observance of the holy canons be maintained, such as the Apostles, so justly praised and worshipped, those eye-witnesses and ministers of God the Word, have delivered down to us, and the holy Fathers have maintained and carried out."[2] And he proceeds to give the force of civil law to the canons concerning the election of bishops and other matters.

In another law he says, "Be it therefore enacted[3] that the force of law be given to the holy canons of the Church which have been set forth or confirmed by the four holy Councils; that is, by the 318 holy Fathers in the Nicene, by the 150 in that of Constantinople, by the first of Ephesus, in which Nestorius was condemned, and by Chalcedon, when Eutyches, together with Nestorius, was put under anathema. For we accept the decrees of these four synods as the Holy Scriptures, and observe their canons as laws.

---

[1] Honestatem quam illis obtenentibus credimus.
[2] *Constitutio*, vi. 48.
[3] 119. *De ecclesiasticis titulis*, p. 940. *Sancimus*. This word in Roman law in the time of Justinian is equivalent to the English formula, "Be it enacted by the Queen's most excellent Majesty, by and with the advice and consent of the Lords spiritual and temporal and the Commons in Parliament assembled, and by the authority of the same". There lies in these two formulæ, expressing the supreme legislative authority, a comparison between the constitution of the lower Roman empire and the medieval constitutions established everywhere by the influence of the Church under guidance of the Popes.

"And, therefore, be it enacted according to their definitions that the most holy Pope of Old Rome is the first of all bishops, and that the most blessed archbishop of Constantinople, New Rome, holds the second place after the holy Apostolic See of Old Rome, but takes precedence of all other bishops."

In the laws just quoted we see three of the most important principles which run through the acts of Justinian. The first is, that the emperor, having the whole commonwealth committed to him by God, is the guardian both of human and divine things in it, which together make up the whole commonwealth; the second is, that there are Two Powers, the human and the divine, both derived from God. The third is, that while the emperor is the direct head of all human things, he guards divine things by accepting the decrees of General Councils as the Holy Scriptures, and by giving to the canons of the Church as descending from the Apostles, "the eye-witnesses and ministers of God the Word," the force of law.

If in these laws we find Church and State greet each other as friends, and offer each other a mutual support, because both aim at one object, and what the holiness of the Church required, advanced no less the peace, the security, and the welfare of the State, so a complete concurrence between them might be shown in all other respects.[1] The State recognised and honoured the whole constitution of the Church as it had been drawn in its first lineaments by the author of the Christian

---

[1] Riffel, 611-12, translated.

religion, as in perfect sequence it had formed itself out of the Church's inmost life, and that in force and purity, because it had been free from the pressure of external laws. The proper position of the Roman bishop as supreme head of the whole Church, the relation of the patriarchs to each other, their privileges over the metropolitans, the close connection of these with their several bishops, were never for a moment unrecognised, because so clear a consciousness of these showed itself in the whole Catholic world, that no change was possible without a general scandal. Thus the laws of Church and State kept pace with each other, when it could not but happen that the ties between patriarch and metropolitan, between metropolitan and bishop, became more stringent, as external increase was followed by decline in inward life and the fervour of faith. Thus the regular course was that the metropolitan examined the election of the bishop by the clergy and people, consecrated him, introduced him to the direction of his charge, and by the *litteræ formatæ* gave him his place in the fabric of the Church. So the metropolitan was consecrated by his patriarch, in whose own election all the bishops of the province, but especially the metropolitans, took part. The metropolitan summoned his bishops, the patriarchs their metropolitans, to the yearly synods. The bishops did not vote without their metropolitan; they took counsel with him, sometimes intrusted him with their votes.[1]

---

[1] See Justinian, *Gloss.* v., directed to the patriarch of Constantinople, Epiphanius. *Epilogus*, p. 48 : Hæc igitur omnia sanctissimi patriarchæ sub se

General laws of the Church, and also imperial edicts, were transmitted first to the patriarchs, and from them to the metropolitans, and from these to the bishops. Bishops might not leave their diocese without permission of the metropolitan, nor the metropolitan without that of the patriarch.[1]

In like manner, we find in Justinian's laws the relation of the bishop to his diocese, and especially to his clergy, recognised as we find it presented by the Church from the beginning, and as the lapse of time had more and more drawn it out. The law's recognition secured it from all attack. The idea that without the bishop there is neither altar, sacrifice, nor sacrament had become, through the spirit of unity which rules the Church, a fact visible to all. The more heresies and divisions exerted their destroying and dissolving power, while the Church went on expanding in bulk, every divine service in private houses was forbidden. Since such assemblies attacked as well the peace and security of the State as the unity of belief, the governors of provinces, as well as the bishops, had most carefully to guard against such acts. Neither in city nor country could a church, a monastery, or an oratory be raised without the bishop's permission. This was made known to all by his consecrating the appointed place in solemn procession, with prayer and singing, by elevation of the

---

constitutis Deo amabilibus metropolitis manifesta faciant, at illi subjectis sibi Deo amabilibus episcopis declarent, et illi monasteriis Dei sub sua ordinatione constitutis cognita faciant, quatenus per omnia Domini cultura maneat undique in eos incorrupta.

[1] Riffel, p. 615, translated.

cross. Without this such building was considered a place where errors lurked and deserters took refuge.[1] In this concurrent action of the laws of Church and State respecting the relation of the bishop to the whole Church and to his own clergy, we never miss the perfect union between the two even as to the smallest particulars. The conclusion is plain that the secular power did not intend to act here on the ground of its own supremacy, or as an exercise of its own majesty. Not only did it issue no new regulations whereby any fresh order should be in the smallest degree introduced: it raised to the condition of its own laws the canons which had long obtained force in the Church, whose binding power was accepted by everyone who respected the Church, as lying in themselves and in the authority from which they proceeded. These it took simply and without addition, and by so taking recognised in them the double character. So, if they were transgressed, a double penalty ensued. The Church's punitive power is contained in its legislative, the recognition of which is an acknowledgment of the former. This the State, not only tacitly but expressly, recognised. And by taking the Church's laws, it not only did not obliterate the character and dignity of that authority, from which they had issued, but it did not change the penalty, nor consider it from another point of view. It remained what it had always been, and from its nature must be, an ecclesiastical punishment. The State only lent its arm, when that was necessary, for its execution. With

[1] Riffel, p. 617.

this, however, it was not content. The Church's life entered too deeply into the secular life. Those who were to carry on the one and sanctify the other stood in the closest connection with the whole State. So it made the canons its own proper laws, and thus attached temporal penalties to their transgression. So we find everywhere the addition that each violation would carry with it not only the divine judgment and arm the Church's hand to punish, but likewise draw down upon it the prescribed penalties from the imperial majesty.

But so far the empire was maintaining by its secular authority the proper laws and institutions of the Church. Justinian went far beyond this.[1] His legislation associated the bishop with the count in the government of cities and provinces. It gave up to him exclusively the superintendence of morality and the protection of moral interests, the control of public works and of prisons. It bestowed on him a large jurisdiction—even more, put under his supervision the conduct of public functionaries in their administration, and conferred on him a preponderating influence on their election. In a word, it by degrees displaced the centre of gravity in political life by investing the episcopate with a large portion of temporal attributions.

To give in detail what is here summed up would involve too large a space. A few specimens must suffice. The bishop in his own spiritual office would have a great regard for widows and orphans.[2] Parents when dying felt secure in recommending children to their protection

---

[1] Kurth, ii. 35. [2] See Riffel, p. 624.

against the avarice of secular judges. Hence the custom had arisen that bishops had to watch over the execution of wills, especially such as were made for benevolent purposes. They could in case of need call in the assistance of the governor. Their higher intelligence and disinterested character were in such general credit that they had no little influence in the drawing up of wills. But the State under Justinian was so far from regarding this with jealousy, that he ordered, if a traveller should die without a will in an inn, the bishop of the place should take possession of the property, either to hand it over to the rightful heirs, or to employ it for pious purposes. If the innkeeper were found guilty of embezzlement, he was to pay thrice the sum to the bishop, who could apply it as he wished. No custom, privilege, or statute was allowed to have force against this. Those who opposed it were made incapable of testing. Down to the sixth century[1] we find no law of the Church touching the testamentary dispositions of Christians. Justinian is the first of whom we know that he entrusted the execution of wills specially to the supervision of bishops. That he did this shows the great trust which he placed in their uprightness.

It was to be expected that bishops should have a special care for the city which was their see.[2] Various laws of Justinian gave them here privileges in which we cannot fail to see the foundation of the later extension of episcopal authority and influence over the whole sphere of secular life. With their clergy and

---

[1] Riffel, p. 625. [2] Ibid., pp. 629-35.

with the chief persons in the city, they took special part in the election of *defensors* and of the other city officers; so also in the appointment of provincial administrators. It was their duty to protect subjects against oppressions from soldiers and exaction of provision, as well as against all excessive claim of taxes and unlawful gifts to imperial officers. A governor on assuming the province was bound to assemble the bishop, the clergy, and the chief people of the capital, that he might lay before them the imperial nomination, and the extent of the duties which he was to fulfil. Thus they were enabled to judge on each occasion whether the representative of the emperor was fulfilling his charge. Magistrates, before entering on office, had to take the prescribed oath before the metropolitan and the chief citizens. The oath itself was an act made before God, and as such under cognisance of the bishop. But special regulations enjoined him to watch over the whole conduct and each particular act of the governor. If general complaints were made of injustice, he was to inform the emperor. If only an individual had suffered wrongs, the bishop was judge between both parties. If sentence was given against the accused, and he refused to make satisfaction, the matter came before the emperor in the last resort. The emperor, if the bishop had decided according to right, condemned his governor to death, because he who should have been the protector of others against wrong had himself committed wrong. If a governor was deposed for maladministration, he was not to quit the province before fifty days, and he could be accused before the bishop for every unjust transaction. Even if he was removed or transferred to another charge, and had left behind him a lawful substitute, the same proceeding took place before the bishop. On this account civil orders also were sent to the bishops to be publicly considered by them, and kept among the church documents, their fulfilment supervised, and violations reported to the emperor. But, to complete this picture, it must be remarked that this supervision was not one-sided. The emperor sent even his ecclesiastical regulations not only through the patriarch of Constantinople to the metropolitans, but through the Prætorian prefect to the governors of provinces. He directed them to support the bishops in their execution, but he likewise enjoined them to report neglect of them to the emperor. Especially they were to watch the execution of imperial decrees upon Church discipline, and monasteries in particular. The rules, so often repeated because so frequently broken, respecting the inalienability of Church property, were to be specially watched, and also the celebration, as prescribed, of yearly synods. But the civil magistrates were only recommended to keep a supervision, which did not extend to the right of official exhortation; far less that they were allowed in any ecclesiastical matter, in which the bishop might be at all in fault, to act upon their own authority, or receive an accusation against him from whomsoever and for whatsoever it might be. But the bishop could act in his quality of judge between a party and the governor himself, if the party had called upon him. Especially, Justinian allowed bishops a decisive influence upon legal proceedings in certain branches. The inspection of forbidden games, public buildings, roads, and bridges, the distribution of corn, was under them. They were to examine the competence of a security. The curators of insane persons took oath before them to fulfil their duty. If a father had named none, the bishop took part in the choice of them; the act was deposited among the church documents. If the children of an insane father wished to marry, the bishop had to determine the dowry and the nuptial donation. In the absence of the proper judge, the bishop of the city could receive complaints from those who had to make a legal demand on another, or to protect themselves from a pledge falling overdue. The proofs of a wrong account could, in the accountant's absence, be made before the bishop, and had legal force. If the ground-lord would not receive the ground-rent, the feoffee should consign it at Constantinople to the Prætorian prefect or the patriarch, in the provinces to the governor, or in his absence to the bishop of the city where the ground-lord who refused to receive it had his domicile. Whoever found no hearing, either in a civil or criminal matter, before the judge of the province, was directed to go to the bishop, who could either call the judge to him, or go in person to the judge, to invite him to do justice to the complainant according to the strict law, in order that the bishop might not be obliged to carry the refusal of justice by appeal to the imperial court.[1] If the judge was not moved by this, the bishop gave the complainant a statement of the whole case for the emperor, and the delinquent had to fear severe penalties, not alone because he had been untrue to his office, but because he did not allow himself, even at the demand of the bishop, to do what, without it, lay in the circle of his duties. But this referring to the bishop was not arbitrary—that is, not one which it lay in the will of the complainant to use or not, but necessary, so that anyone who appealed to the imperial court without this endeavour incurred, whether his complaint was founded or not, the same punishment as the judge who refused to give a decision at the bishop's request. Even if the complainant only suspected the judge, he was bound to apply to the bishop to join the judge in examining the matter, and to bring it to a strict legal issue. In the face of such honourable confidence which was placed in the bishops, and which was also justified in general by a happy result, we ought not to be surprised if either the emperor himself or inferior magistrates committed to them the termination of entangled processes, in which they exercised just such a jurisdiction as may either in general be exercised by delegates, or was committed to them for the special occasion.

The emperor[2] in his legislation left no part of the

---

[1] See St. Gregory, *Epis.*, x. 51 (vol. ii. 1080), where he writes to the ex-consul Leontius, in Sicily, who had beaten with rods the ex-prefect Libertinus: "Si mihi constare potuisset quia justas causas de suis rationibus haberent, et prius per epistolas vos pulsare habui; et si auditus minime fuissem, serenissimo Domino Imperatori suggererem".

[2] Riffel, p. 635.

Church's discipline unregarded. His purpose was in all respects to make the State Christian; and he considered no part of divine and human things, whether it were dogma or conduct,—which, together, made up the Church's life,—withdrawn from his care and guardianship. Observances which had begun in custom, and gradually been drawn out definitely and enacted in canons, he took into his *Digest*, not with the intention of giving them greater inward force or stronger grounds as duties, but to show the unity of his own effort with that of the Church. He willingly put the imperial stamp on her salutary regulations. He showed his readiness to help her with external force wherever the inviolable sanctity of her laws seemed to be threatened by the opposition of individuals. In this he recognised the unchangeable order which is so deeply rooted in the nature both of Church and State, that order which is the greatest security for the wellbeing and prosperity of both. And the Church in the course of her long life had hitherto almost universally maintained this order; always, at least, in principle. If it was anywhere transgressed, it was either because the secular power was acting under special commission and approval of the Church, or, if that power acted without such approval, it met with open contradiction whereby not only the illegality of the particular action was marked, but the principle of the Church's freedom and independence was preserved.

There is a passage in the address of the eastern bishops to Tarasius, patriarch of Constantinople, quoted in the Second Nicene Council of 789,[1] the Seventh General, which cites the words of Justinian given above in one of his laws. The bishops say in their own character—and they are bishops who describe themselves "as sitting in darkness and the shadow of death, that is, of the Arabian impiety"—"It is the priesthood which sanctifies the empire and forms its basis; it is the empire which strengthens and supports the priesthood. Concerning these, a wise king, most blessed among holy princes, said: The greatest gift of God to men is the priestly and the imperial power, the one ordering and administering divine things, the other ruling human things by upright laws."

If we considered the principles of Justinian alone as exhibited in his legislation, without regard to his conduct, we might, like the eastern bishops, take these words as the motto of his reign and the key to his acts as legislator. Indeed, it may be said that this legislation cannot be understood except by presupposing throughout the cordiality of the alliance between the Two Powers. In the election and the lives of bishops, in the discipline of religious houses, in the strict observance of the celibate life which has been assumed with full consent of the will by clergy and by monks, the emperor is as strict in his laws as the Church in her canons. The ruler of the State, who makes laws with a single word of his own mouth, who commands all the armies of the State, who bestows all its offices, who is, in truth, the autocrat, the impersonated commonwealth,

[1] Mansi, xii. 1130.

shows not a particle of jealousy towards the Church as Church. He enjoins the strict observance of her canons in the fullest conviction that the end which she aims at as Church is the end which he also desires as emperor; that the good life of her bishops and priests is essential for the good of society in general; that the perfect orthodoxy of her creed is the dearest possession, the pillar and safeguard, of his own government. Heresy and schism are, in his sight, the greatest crimes against the State, as they are the greatest sins against the Church and against God. In the course of the two hundred years from Constantine to Justinian the Roman State, as understood by the Illyrian peasant who ruled it for thirty-eight years, had intertwined itself as closely with the Catholic Church as ever it had with Cicero's "immortal gods" in the time of Augustus, or Trajan, or Decius. It was the special pride and glory of Justinian to maintain intact this alliance as the palladium of the empire. And, therefore, his legislation touched every part of the ecclesiastical government, every dogma of the Church's creed, and only on account of this alliance did the Church acquiesce in such a legislation. I suppose that no greater contradiction can ever be conceived than that which exists between the mind of Justinian and the mind which now, and for a long time, has directed the nations of Europe, so far as their governments are concerned in their attitude towards the Church of God. In Europe are nations which are nurtured upon heresy and schism, whether as the basis of the original rebellion which severed them from the communion of the Church or as the outcome of " Freethought" in their subsequent evolution through centuries of speculation unbridled by spiritual authority; nations, again, bisected by pure infidelity, or struggling with the joint forces of heresy and infidelity which strive to overthrow constitutions originally Catholic in all their structure. In one empire alone the attitude of Constantine and Justinian towards the Church is still maintained. It is that wherein the emperor rules with an amplitude of authority such as Constantine and Justinian held, whose successor he claims to be; where, also, an imperial aide-de-camp, booted and spurred, sits at the council board of a synod called holy, and is by far the most important member of it, for nothing can pass without his sanction—a synod which rules the bishops, being itself nothing but a ministry of the State, drawing, like the council of the empire, its jurisdiction from the emperor.

Justinian was a true successor of the great Theodosius in so far as he upheld orthodoxy, and endeavoured to unite all his subjects in one belief and one centre of unity. The greatest of the Roman emperors had for their first and chief motive, in upholding this first principle of imperial policy, the conviction that thus only they could hope to maintain the peace and security of the empire. Schism in the Church betokened rebellion in the State. In the fourth century heresy had driven the empire to the very brink of destruction. Besides this, all the populations converted from heathendom were accustomed to see a complete harmony between religion and the State, which appeared almost blent into one. Again, we must not forget that at this time the Christian religion had been lately accepted distinctly as a divine institution, and that it embraced the whole man with a plenitude of power which the indifference and division of our own times hardly allow us to conceive. Those who would realise this grasp of the Christian faith, tranforming and exalting the whole being, may reach a faint perception of it by reading the great Fathers of the fourth and fifth centuries—St. Basil, St. Ambrose, St. Chrysostom, St. Jerome, St. Augustine, and St. Leo. They were not in danger of taking the moral corruption of an effete civilisation for the Christian faith. Again, the emperors, living in the midst of this immense intellectual and moral power—for instance, Justinian himself practising in a court the austerities of a monastery—recognised the confession of the same faith as the strongest band which united subjects with their prince. They thought that those who were not united with them in belief could not serve them with perfect love and fidelity. And, lastly, they hoped that their own zeal in maintaining the Church's unity unimpaired would make them worthier of the divine favour, and give success to all their undertakings. Let us take the words of Theodosius, one of the greatest and best among them, to his colleague the younger Valentinian, who up to the time of his mother Justina's death had been unjust to the Catholic cause and favoured the Arian heresy : " The imperial dignity is supported, not by arms, but by the justice of the cause. Emperors who feared God have won victories without armies, have subdued enemies and made them tributary, and have escaped all dangers. So Constantine the Great overcame the tyrant Licinius in a sea-fight. So thy father (the first Valentinian) succeeded in protecting his realm from its enemies, won mighty victories, and destroyed many barbarians. On the contrary, thy uncle Valens polluted churches by the murder of saints and the banishing of priests. Hence by guidance of Divine Providence he was besieged by the Goths, and found his death in the flames. It is true that he who has not unjustly expelled thee does not worship Christ aright. But thy perverse belief has given this opportunity to Maximus. If we do not return to Christ, how can we call upon His aid in the struggle?" The following emperors were of the same judgment: so that they attached to each decree which concerned ecclesiastical matters the motive of meriting thereby God's approval, since they not only took pains to please Him, but also led their subjects to do so. We employ, says Justinian, every care upon the holy churches, because we believe that our empire will be maintained, and the commonwealth protected by the favour of God, but likewise to save our own souls and the souls of all our subjects.

Justinian likewise would have a keen remembrance of the degradation from which his uncle had restored the empire. None knew better than he how the ignoble reigns of the usurper Basiliscus, of Zeno, and of Anastasius, by perpetual tampering with heresy and ruthless persecution of the orthodox, had well-nigh broken that empire to pieces. Had he not thrown all his energy, as the leading spirit of his uncle's realm, into that great submission to Pope Hormisdas which rendered its beginning illustrious?

Nevertheless a dark blot lies upon the name and memory of Justinian. He was not only successor of the great Theodosius in his ardent zeal for the Church's doctrine and unity, but likewise of Constantine, when he sullied his greatness and risked all the success of his former life by falling into the hands of the Nicomedian Eusebius.

The vast event by which the Christian Church had become a ruling power in the commonwealth had affected from that time forth the whole being of Church and State. Christian emperors had come to see in bishops the Fathers and Princes of such a Church, consecrated by God to that office, not appointed by men.[1] As such they had honoured them, committed to their wisdom and guidance the salvation of their own souls, and the weal itself of the commonwealth; not hindered them in the performance of their duties, not hampered them by restrictive laws. Rather they had protected them by external force from hindrance when invited thus to show their protection as heads of the State. Circumstances led them on to a more immediate entrance into the Church's special domain, and the things which happened in that domain

---

[1] Riffel, 562.

led to this their entrance. It kept even pace with the developments and disturbances caused by heresy therein.

Christ had committed to the whole episcopate, under the guidance of the Holy Spirit, the task of spreading the seed of Christian doctrine over the earth, of watching its growth, of eradicating the false seed sown in night-time by the enemy. In proportion as the empire's head took part in this work, his influence on the episcopate could not but increase. If his participation was confined within its due limits, if the temporal ruler hedged the Church round from irruption of external power, if he rooted the tares out of her field only to clear her enclosure, his relation to the bishops remained merely external. But if he went on himself to lay down the limit of the Church's domain, or even if he only took an active part in such limitation; if he made himself the judge what was wheat and what was tares, in so doing he had won an influence on the bishops which did not belong to him. Then Church and State ran a danger of seeing their respective limits confused. Thus the relation of the bishops to the ruler of the State became then, and remains always, an unfailing standard of the Church's freedom and independence.

Now, striking and peremptory as the eastern submission to Pope Hormisdas was, in which Justinian, then a man of thirty-six, had taken large part; clear and unambiguous as in his legislation appears the recognition of the Two Powers, sacerdotal and imperial, which make together the joint foundation of the State, and are a necessity of its wellbeing; distinct, likewise, as is the imperial proclamation of the Pope as the first of all bishops in his laws, his letters, confirmed by his reception of the Popes Agapetus and Vigilius in his own capital city; frank and unembarrassed as his acknowledgment of St. Peter's successors, yet, when he had reached the mature age of seventy, and was lord by conquest of Rome reduced to absolute impotence, and of Italy as a subject province, his treatment of the first bishop, in the person of Vigilius, was a contradiction of his own laws as to the two domains of divine and human things. He passed beyond the limits which marked the boundaries of the two powers. He made himself the supreme judge of doctrine. He convoked a General Council without the Pope's assent; he terminated it without his sanction; he treated the Pope as a prisoner for resisting such action. It is true that St. Peter's successor—and this with a stain upon him which no successor of St. Peter had worn before him—escaped with St. Peter's life in him unimpaired; but so far as the action of Justinian went it was unfilial, inconsistent with his own laws, perilous in the extreme to the Church, dishonouring to the whole episcopate. The divine protection guarded Vigilius—that Vigilius whom an imperious woman had put upon the seat of a lawful living Pope—from sacrifice of the authority to which, on the martyrdom of his predecessor, he succeeded. He died at Syracuse, and St. Peter lived after him undiminished in the great St. Gregory. The names mean the same, the one in Latin, the other in Greek; but no successor ever took on himself the blighted name of Vigilius, while many of the greatest among the Popes have chosen for themselves the name of Gregory, and one at least of the sixteen has equalled the glory of the first.

In judging the conduct of Justinian, both in treatment of persons and in dealing with doctrine, we cannot fail to see that the imperial duty of protection passed into the imperial lust for mastery. If his treatment of Vigilius, whom he acknowledged in the clearest terms as Pope, was scandalous and cruel, still worse, if possible, was the assumption of a right to interpret and to define the Church's doctrine for the Church. The usurper Basiliscus had been the first to issue an imperial decree on doctrine. This was in favour of heresy. He was followed in this by the legitimate emperors Zeno and Anastasius, also in favour of heresy. On the contrary,[1] the edicts of Justinian were generally in conformity with the decisions of the Church: generally occasioned by bishops, often drawn up by them. But in the council called by him at Constantinople in 553, he issued decrees on doctrines which only the Church could decide. In doing this he infringed her liberty as grossly as the three whose unlawful act he was imitating. The whole effect of his reign was that State despotism in Church matters lowered the dignity of the spiritual power. The dependence of his

[1] Photius, p. 155.

bishops on the court became greater and greater. The emperor's will became law in the things of the Church. He persecuted Vigilius: he deposed his own patriarch Eutychius. His example, as that of the most distinguished Byzantine monarch, told with great force upon his successors, for the persecution of future Popes and the deposition of future patriarchs.

The Italy which he had won at the cost of its ruin as to temporal wellbeing was, after his death in 565, speedily lost as to its greater portion, and the Romans [1] of the East did little more for it. The Rome which he had reduced almost to a solitude, and ruled through a prefect with absolute power, escaped in the end from the most cruel and heartless despotism inflicted by a distant master on a province at once plundered and neglected. His own eastern provinces suffered terribly from barbarian inroads, and the end of the thirty-seven years' domination, which had seemed a resurrection at the beginning, showed the mighty eastern empire from day to day declining, the western bishops under the action of the Pope more and more exerting an independence which the East could not prevent, the patriarch of Constantinople more and more advancing as the agent of the imperial will in dealing with eastern bishops. What the See of St. Peter was at the end of the sixth century it remains to see in the pontificate of the first Gregory, who shares with the first Leo the double title of Great and Saint.

[1] Photius, 173.

CHAPTER V.

ST. GREGORY THE GREAT.

> "The banner of the Church is ever flying!
> Less than a storm avails not to unfold
> The Cross emblazoned there in massive gold:
> Away with doubts and sadness, tears and sighing!
> It is by faith, by patience, and by dying
> That we must conquer, as our sires of old."
> —AUBREY DE VERE, "St. Peter's Chains".

THE historian,[1] who has carefully followed the fortunes of Rome as a city during a thousand years, describes it as beginning a new life from the time when Narses, in the year 552, came to reside there as imperial prefect and representative of the absent eastern lord Justinian. Narses so ruled for fifteen years, but when he was recalled there ensued a long time of terrible distress and anxiety—a time of temporal servitude, but one also of spiritual expansion. The complete ruin of Rome as a secular city, the overthrow of all that ancient world of which Rome was the centre and capital, had been effected in the struggle ended by the extinction of the Gothic kingdom. By degrees the laws, the monuments, the very recollections of what had been, passed away. The heathen temples ceased to be preserved as public monuments. The Capitol, on its desolate hill, lifted into the still air its fairy world of pillars in a grave-like

---
[1] See Gregorovius, ii. 3, 4.

silence, startled only by the owl's night cry. The huge palace of the Cæsars still occupied the Palatine in unbroken greatness, a labyrinth of empty halls yet resplendent with the finest marbles, here and there still covered with gold-embroidered tapestry. But it was falling to pieces like a fortress deserted by its occupants. In some small corner of its vast spaces there might still be seen a Byzantine prefect, an eunuch from the court of the eastern despot, or a semi-Asiatic general, with secretaries, servants, and guards. The splendid forums built by Cæsar after Cæsar, each a homage paid by the ruler of the day to the Roman people, whom he fed and feared, became pale with age. Their history clung round them like a fable. The massive blocks of Pompey's theatre showed need of repairs, which were not given. The circus maximus, where the last and dearest of Roman pleasures—the chariot races—were no longer celebrated, stretched its long lines beneath the imperial palace covered with dust and overgrown with grass. The colossal amphitheatre of Titus still reared its circle perfect, but stripped of its decorations. The gigantic baths, fed by no aqueduct since the ruin wrought by Vitiges the Goth, rose like fallen cities in a wilderness. Ivy began to creep over them. The costly marble mantle of their walls dropped away in pieces or was plundered for use. The Mosaic pavements split. There were still in those beautiful chambers seats of bright or dark marble, baths of porphyry or Oriental alabaster. But these found their way by degrees to churches. They served for episcopal chairs, or to receive the bones of a saint, or to become baptismal fonts. Yet not a few remained in their desolation till the walls dropped down upon them, or the dust covered them for centuries. In course of time the rain perforated the uncared-for vaultings of these shady galleries. Having served for refuge to the thief, the coiner, or the assassin, they became like dripping grottoes.

Thus stood the temples, triumphal arches, pillars, and statues before the eyes of a young Roman noble, one out of the few patrician families still surviving. These were the sights with which St. Gregory, who claimed kindred with the Anician race, was familiar from his boyhood, so that the desolation of Jerusalem rose before his mind as the state of his own Rome pressed on his eyes and seared his heart.

This skeleton of a city was scarcely inhabited by the remnant of a people, decimated by hunger and pestilence, and in perpetual fear to see its ill-defended gates broken into by Lombard savages. The walls of Aurelian, half demolished by Totila and hurriedly repaired by Belisarius, alone saved it year after year from the horrors which fell upon captured cities; and would not have saved it but for the indomitable spirit, the perpetual wisdom, foresight, and courage of a son who had been exalted to the Chair of Peter.

While Old Rome lay thus, the shadow of its former self, bereft of all political power, looking to the imperial exarch at Ravenna for its temporal rule, in danger moreover of inundation from its own Tiber, whose banks were no longer maintained with unremitting care, New Rome beside the Bosporus rioted in all the pomp and circumstance of a court still the head of a vast empire. The tributes of all the East, of numberless cities in Asia Minor, in Syria, in Egypt, were still borne unceasingly within its walls, which rose as an impregnable fortress between Europe and Asia. Its emperor still thought himself the lord of the world; its bishop assumed the title of Ecumenical Patriarch. Both emperor and bishop cast but a disdainful glance on the widowed rival which threatened to sink into the grave of waters brought down by her own river. Constantinople could raise and pay armies from all the races of the North and East. A single imperial regiment was quartered at Rome, which, being ill-paid, became disaffected and neglectful of its charge, and could not be counted upon by the Pope for vigorous defence against the ever-pressing danger of a Lombard inroad.

So began the Church's Rome.[1] Enslaved politically to Byzantium, wherein the so-called Roman State, with Greek subtlety, carried on the principles of the old heathen government and practised a remorseless despotism, the city of the ancient Cæsars and the people they fed on "bread and games" ceased to exist, and was changed into the holy city, whose life was the Chair of Peter. From the time of Narses, during all the two hundred years of Lombard assault and Byzantine neglect and exaction, the Pope alone, watchful and unceasingly active, carried out the fabric of the Roman hierarchy.[2] Its gradual increase, its springing up out of the dust of the

[1] Gregorovius, ii. 6. [2] *Ibid.*, ii. 5, literal.

old Roman State under the most difficult circumstances, will ever claim the astonishment of the after-world as the greatest transformation to be found in history.

Let us approach the secret of this transformation in the person of the man who best represents it.

Gregory was born about the year 540, and so was witness from his childhood of the intense misery and special degradation of Rome produced by the Gothic war. He was himself the son of Gordian, a man of senatorial rank, from whom he inherited great landed property. Through him he was the great grandson of that illustrious Pope Felix III., whom we have seen resist with success the insolence of Acacius and the despotism of Zeno. Gregory had therefore a doubly noble inheritance—that of a true Roman noble's spirit, and that of the Church's championship. His paternal house stood on that well-known slope of the Cœlian hill, opposite the imperial palace on the Palatine, from which in after-time he sent forth St. Augustine with the monks his brethren to be the Apostle of paganised England. He founded six monasteries in Sicily upon his property, and changed his father's palace into a seventh, in which he followed the Benedictine Rule. In early manhood he had been prætor or prefect of the city, being probably the most eminent of all its citizens in wealth and rank. But his mother St. Silvia, a woman of fervent piety, had educated him with great care. He turned from the secular to the religious life, following perhaps her example, since on the death of his father she became a nun. He was a monk on the

Cœlian hill when Pope Benedict in the year 577 named him seventh deacon of the Roman Church. Pope Pelagius II. sent him as nuncio to Constantinople, an office equally difficult and honourable. The emperor Tiberius was then reigning, with whom he became intimate, and with his successor Mauritius. Gregory dwelt in the imperial palace, with some monks of his own monastery whom he had brought with him, pursuing the Rule in all pious observances, winning also the esteem and friendship of many distinguished men, and making himself fully acquainted with the mechanism of the eastern court. He also delivered the patriarch Eutychius from a false Origenistic notion, that the bodies of the blessed after the resurrection were not glorified, but lost their quality as bodies.[1] There also he became warmly attached to St. Leander, who afterwards, as archbishop of Seville, greatly helped him in recovering Spain from Arianism to the Catholic faith. The charge of Pope Pelagius to his nuncio Gregory throws a vivid light upon the condition of Rome at the time. His instructions ran: "Lay before our lord the emperor that no words can express the calamities brought upon us by the perfidy of the Lombards, breaking their own engagements. Our brother Sebastian, whom we send to you, has promised to describe to him the necessities and dangers of all Italy. Join him in that entreaty to succour us, for the commonwealth is in such distress, that unless God inspire him to show us his servants the mercy of his natural disposition, and move

---

[1] Nirschl, iii. 534.

him to give us a single *Magister militum* and a single *Dux*, we are utterly destitute, for Rome and its neighbourhood are specially defenceless. The exarch writes that he can give us no help, for he has not force enough to guard Ravenna. Therefore, may God command the emperor quickly to succour us, before the army of that most wicked nation take the places still remaining to us."[1]

Gregory returned from Constantinople in 585, and lived as one of the seven deacons on the Cœlian hill, when, on 8th February, 590, Pope Pelagius died of the pestilence, and Gregory was unanimously chosen to succeed him.

It was a moment of the greatest depression. The Tiber had in the winter overflowed a large portion of the city. The destruction wrought had been followed by a terrible plague. Gregory strove to escape the charge put upon him, and besought the emperor not to confirm his election. In the meantime, the clergy and people urged upon him the provisional exercise of the episcopal charge. As such he ordered a sevenfold procession to entreat the cessation of the plague. The clergy of Rome, the abbots, the abbesses with their nuns, the children, the laymen, the widows, and the married women, each company separately arranged, were to start from seven different churches, and to close their pilgrimage together at the basilica of St. Maria Maggiore.

During the procession itself eighty victims to the plague fell dead. But as Gregory was passing over the bridge of St. Peter's, a heavenly vision consoled them in the midst of their litanies. The archangel Michael

---

[1] Third letter of Pelagius II.; Mansi ix., p. 889: *Nefandissima gens*.

was seen over the tomb of Hadrian, sheathing his flaming sword in token that the pestilence was to cease. Gregory heard the angelic antiphon from heavenly voices—*Regina Cœli, lætare,* and added himself the concluding verse—*Ora pro nobis Deum, alleluia.*

The assent of the emperor Mauritius arriving from Constantinople about six months after his election compelled Gregory to become Pope. At first, indeed, he disguised himself and took to flight, and hid himself in the woods.[1] The people fasted and prayed three days for his discovery. He was found, and then permitted himself to be taken back to Rome, where he was received with great rejoicing. He was led, according to custom, to the "Confession" of St. Peter, where he made his profession of faith. He was then consecrated, the 3rd September, 590. Nor can any words but his own adequately express his feelings, together with the character of the time in which he lived. With heavy heart he approached the burden laid upon him. Neither then nor ever after did he deceive himself as to the gravity of the situation. "Since," are his words, "I submitted the shoulders of my spirit to this burden of the episcopal office, I can no longer collect my soul, distracted as it is on so many sides. At one time I have to consider the affairs of churches and monasteries, often taking into account the lives and actions of individuals. At another time I have to represent my fellow-citizens in their affairs. Again, I

---

[1] Attested by St. Gregory of Tours, who heard it from a deacon of his church then at Rome.

have to groan over the swords of barbarians advancing to storm us, and to dread the wolves which lie in wait for a flock huddled together in fear. Then, again, I must charge myself with the care of public affairs, to provide means even for those to whom the maintenance of order is entrusted, or I must patiently endure certain depredators, or take precautions against them, that tranquillity be not disturbed." In another place he says: "Daily I feel what fulness of peace I have lost, to what fulness of cares I have been exalted. If you love me, weep for me, since so many temporal businesses press on me that I seem as if this dignity had almost excluded me from the love of God. Not of the Romans only am I bishop, but bishop of the Lombards, whose right is the right of the sword, whose favour is punishment. The billows of the world so surge upon me, that I despair of steering into harbour the frail vessel entrusted to me by God, while my hand holds the helm amid a thousand storms." Again, in his synodical letter[1] announcing his accession to the patriarchs, he says: "Especially, whoever bears the title of Pastor in this place is grievously occupied by external cares, so that he is often in doubt whether he is executing the work of a Pastor or that of an earthly lord". Thus thirteen hundred years ago spoke the Pope. Does his language in the nineteenth century differ much from his language in the sixth? Shortly after his accession, preaching to his people in St. Peter's, he said:[2] "Where, I pray you, is any delight to be found in

---

[1] *Ep.* i. 25, p. 514. [2] *Homily* xviii. *on Ezechiel*, tom. i. 1374.

this world? Mourning meets us everywhere; groans surround us. Ruined cities, fortresses overthrown, lands laid waste, the earth reduced to a desert. The fields have none to till them.' There is scarcely a dweller in the cities. Yet even these poor remnants of the human race are smitten daily and without ceasing. The scourge of heaven's justice strikes without end, because even under its strokes our bad actions are not corrected. We see men led into captivity, beheaded, slain before our eyes. What pleasure, then, does life retain, my brethren? If yet we are fond of such a world, it is not joys but wounds which we love. We see the condition of that Rome which anon seemed to be mistress of the world: worn down by sorrows which have no measure, desolate of inhabitants, assaulted by enemies, filled with ruins. We see in it fulfilled what long ago our prophet said against Samaria: 'Set on a vessel; set it on, I say, and put water into it. Heap together into it the pieces thereof.' And then: 'The seething of it is boiling hot; and the bones thereof are thoroughly sodden in the midst thereof'. And further: 'Heap together the bones, which I will burn with fire: the flesh shall be consumed, and the whole composition shall be sodden, and the bones shall be consumed. Then set it empty upon burning coals, that it may be hot, and the brass thereof may be melted.' Now the vessel was set on when our city was founded. The water was put into it and the pieces heaped together, when there was a confluence of peoples to it from all sides. Like boiling water they bubbled up with the world's actions; like bits of flesh they were boiled in their own heat. He says well, 'The seething of it is boiling hot, and the bones thereof are thoroughly sodden in the midst thereof'. For great, indeed, in it at first was the heat of secular glory; but presently the glory itself and those who followed it burnt out. Bones mean the powerful of the world; flesh its various peoples: as bones support flesh, so the powerful of the world rule the weakness of the masses. But now, behold, all the the powerful of this world have been taken from it. The bones, then, are thoroughly sodden. The peoples are gone; the flesh, then, is boiled up. There follows then: 'Heap together the bones, which I will burn with fire; the flesh shall be consumed, and the whole composition shall be sodden; and the bones shall waste away'. For where is the senate? where any longer a people? The bones are wasted, the flesh consumed; all pride of secular dignities is perished out of it. The whole composition is sodden. Yet every day the sword, every day innumerable sorrows press upon us, the poor remaining remnant. So, then, this also applies: 'Set it empty upon burning coals'. For since there is no senate, since the people has died out, and yet sorrow and suffering are multiplied day by day on the few that remain, Rome is empty, and yet it burns. We apply this to men, but we see the very structures destroyed by the multiplication of ruins. So that he adds, upon the empty city, 'Burn it and melt its brass'. For it is come to the vessel itself being destroyed, in which before both flesh and bones were consumed. For when the dwellers have fallen away even the walls fall. But where are those who once rejoiced in its glory? Where is their pomp and pride, and those ecstasies of frequent transport?

"In Rome are fulfilled the prophet's words against Niniveh: 'Where is the dwelling of the lions, and the feeding-place of the young lions?'[1] Were not its commanders and its princes lions who overran the whole world, and ravened, and slaughtered the prey? Here the young lions found their feeding-place, because the boyhood, the youth, the flower of manhood, from generation to generation, flocked hither, when they sought to get on in the world. Now Rome is desolate, worn down, full of sorrows. No one comes to it to get on in the world; no man of power or violence remains to raven on the prey. Then may we say, 'Where is the dwelling of the lions, and the feeding-place of the young lions?' Upon it has fallen the lot of Judea, foretold by the prophet: 'Enlarge thy baldness as the eagle'.[2] For man is wont to be bald upon the head alone; but the eagle's baldness is over all his body. When very old, his plumes and feathers fall from his whole body. The city which has lost its inhabitants, in losing its feathers, has enlarged its baldness as the eagle. Shrunk also are its wings, with which it used to fly to the prey, for all its men of might, by whom it ravened, are extinguished."

We may here contrast the language concerning the

[1] Nahum ii. 11.   [2] Micheas i. 16.

Rome which lay before their eyes of the two Popes St. Leo and St. Gregory. They spoke with an interval between them of 140 years. The first spoke still of the actual queen of the world, of the secular empire subdued and inherited by the spiritual. The feathers of Leo's eagle shone to him with celestial light; the talons of the royal bird traversed the earth not to raven, but to feed a conquered world with Christian doctrine. St. Gregory speaks of the eagle as bald; but we shall see that he who day by day guarded the gates of defenceless Rome against the Lombard spoiler, barbarian also and heretic, fed no less the ends of the earth with Christian doctrine. It was he who brought the *Ultima Thule*, and its inhabitants the *penitus toto divisos orbe Britannos* again under the yoke of Christ, and taught the sea-kings humanity.

A little later St. Gregory closed his exposition of the prophet Ezechiel in St. Peter's with these sorrowful words: "So far, dear brethren, by the gift of God, we have searched out hidden meanings for you. Let no man blame me if I close them here, because, as you all witness, our sufferings have grown enormous. On every side we are encircled with swords: on every side we are in imminent peril of death. Some return to us maimed of their hands; of others we hear that they are captured; of others, again, that they are slain. My tongue can no longer expound, when my spirit is weary of my life. Let no one ask me to unfold the Scriptures; for my harp is turned to mourning, and my voice to the cry of the weeper. The eye of my heart no longer keeps its watch in the discussion of mysteries; my soul droops for weariness. Study has lost its charm for me. I have forgotten to eat my bread for the voice of my groaning. How can one who is not allowed to live take pleasure in the mystical sense of Scripture? How can one whose daily chalice is bitterness present sweets for others to drink? What remains for us but while we weep to give thanks for the strokes of the scourge which we suffer for our iniquities. Our Creator is become our Father by the Spirit of adoption whom He has given to us: sometimes He feeds His sons with bread; sometimes He corrects them with the scourge; because He schools us by sorrows and by gifts for the unending inheritance."[1]

This was the Rome in which Gregory ruled as Pope for fourteen years, since he saw the archangel's sword sheathed over the castle of St. Angelo, into which name the pagan mausoleum was baptised. Pestilence in the city, where the remnant of a people wandered disconsolate by the mighty halls and vast spaces of the old emperors—swords of pagan or Arian barbarians all round the patched-up walls of Aurelian. City after city through the hapless Italy reported as plundered or ruined by the Lombard devastation. Presently the trials of a sick-bed and frequent attacks of gout were added to his daily tale of sorrows. In the last years of Gregory it came to pass that the universal Church was governed from the sick-bed of one worn down, not by years—for he died at sixty-four—but by sufferings of body and mind. The

[1] End of the *Homilies on Ezechiel*, tom. i. 1430.

prisoner of the Lombards had to struggle perpetually with the spirit of Byzantine despotism and the aggressive arrogance of a prelate whom successive eastern sovereigns had nursed from a suffragan of Heraclea to be the claimant of an ecumenical patriarchate. Yet the eyes of Gregory were bent likewise on the northern conquerors who had seized the provinces of the West. Before he was Pope he had observed in the slave-market of Rome the fair-haired Angles whom he would fain make angels; when Pope he sent forth from his father's house, which he had given to the great Father Benedict, those who were to carry the banner of that father into the isle lost to Christ. In that island he appointed the primate of Canterbury, and designed the primate of York. Through St. Leander and St. Isidore, and the martyr St. Hermenegild, he recovered Spain from the Arian blight; through the queen Theodelinda he made some impression upon Lombard cruelty and misbelief; through the Frankish monarchy he won back France from dissolution and heresy. As he saw the palaces around him deserted, and the broken aqueducts mourn over their intercepted streams in a wasted Campagna, and the glory of Trajan's forum become paler day by day, he thought that the end of the world was coming —and so thinking and so saying, he founded Christendom. In Rome itself, the almsgivers whom he had organised traversed the streets daily, carrying food to the hungry, medicine and medical aid to the sick. Every month he allotted portions of corn, wine, oil, cheese, fish, vegetables. The Church seemed to be the general provider. Every day he fed at his table twelve poor pilgrims, and served them himself. The nuns who took refuge in Rome, from the destruction of their monasteries by the Lombards, amounted to three thousand, whom Gregory supported, especially during the severe winter of 597. He wrote to the sister of the emperor Mauritius: "To their prayers and tears and fasts Rome owes its delivery from the sword of the Lombards".[1] Other cities also he saved, and so he distributed the vast patrimony of the Roman Church in Southern Italy, Sicily, Africa, France, Illyricum, with such wisdom and so beneficent a mercy, that historians trace to him the beginning of that temporal sovereignty which two hundred years after him the Popes were to take in change for the cruel abandonment, paired with incessant exaction, of Byzantine despotism; and the most loyal of subjects were called to be the most beneficent of sovereigns; and the people who had found them fathers from age to age rejoiced to see the fathership united with kingship.

What had happened to the Italy recovered by the arms of Belisarius and Narses, to the unity of the Roman empire, which caused the calamitous state described by Gregory?

Both Belisarius and Narses had enrolled a multifarious host of adventurers under the banner which professed to deliver Rome and Italy from the Gothic occupation. Narses especially had awakened the greed of the Lombards by the sight of Italy's fair lands.

[1] Quoted by Reumont, ii. 90.

Scarcely had he ceased to govern Rome, in 567, when the effect of this became visible. What Alaric, what Odoacer, what Theodorick, had done, Alboin did with yet more terrible results; and the fourth captivity which Nova Roma had prepared for her mother, become in her mind a hated rival, was the hardest, the longest, the most destructive of all. It is doubtful whether the retort of the eunuch Narses to the empress Sophia, when she recalled him from his government to ply, as she said, the spindle, that he would spin for her such a thread as in her life she would not disentangle, is authentic, but it undoubtedly presents historic truth. Whether or not Narses called the Lombards into Italy, their king Alboin came from Pannonia over the Carnian Alps into the plain which has ever since borne their name; and this was in the next year—568—to the recal of Narses. The Goth and the Herules had worked much woe and wrought great destruction; but the Goths compared to the Lombards were as knights compared to villains. The Lombards, inferior to them by far in strength both of body and of mind, this rudest of Teuton races seemed incapable of receiving culture. It had, moreover, fewer elements in it capable of being worked into the stable order of a state. In belief it was partly Arian and partly pagan. It had also a mixture of Sarmatian blood. When they broke into Italy, the cities of that land, however wasted and depopulated through Attila and the Gothic wars, yet retained their Roman form, yet were full of ancient monuments, splendid still in desolation. Now, one after another fell under the sword of those barbarians. Milan surrendered to Alboin in the autumn of 569, and after three years' siege he entered as conqueror into Theodorick's palace in Pavia. Only Rome, Ravenna, and the cities of the coast still carried the imperial flag. The Romans themselves regarded as a marvel the maintenance of their scarcely defended city. Alboin aimed at making the palace of the Cæsars his royal residence. His warriors advanced with terrible devastation from Spoleto to the very walls of Rome in the time of Pope John III., who died, after nearly thirteen years' government, the 13th July, 573.

Rome was then so severely pressed that the See of Peter remained more than a year unfilled; for the Lombards were encamped before Rome, and hindered communication with Byzantium, whence Benedict I., the newly-elected Pope, had to wait for the imperial confirmation. The *Book of the Popes* recites that during his four years' government the Lombards overran all Italy, and that pestilence and hunger consumed her people. Rome, also, was visited by both. The emperor Tiberius tried to succour it by sending corn from Egypt to the harbour Porto.

Alboin had been murdered, and Kleph had succeeded him, on whose death, in 575, the Lombards fell into anarchy, and were divided into thirty-six dukes, and Faroald, the first duke of Spoleto, held Rome besieged when Benedict I. died, in 578; and so his successor, Pelagius II., a Roman of Gothic descent, was consecrated without the emperor's confirmation. The beleaguered Pope sent a cry of distress by an embassy to the eastern emperor, together with a gift of 3000 pounds' weight of gold from the impoverished city. But the emperor, engaged in a Persian war, could only send insufficient troops to Ravenna, more precious to him than Rome, declined the Roman gold, and advised to corrupt with it the Lombard commanders. Zoto, the Lombard duke of Beneventum, returning from Rome, which had ransomed itself, destroyed St. Benedict's monastery of Monte Cassino, in 580. The monks escaped to Rome, carrying with them the Saint's autograph of his Rule. Pope Pelagius II. received them in the Lateran basilica. There they founded the first Benedictine monastery in Rome. They named it after St. John the Baptist and St. John the Evangelist, and so Constantine's basilica, or the Church of the Saviour, became in after-times St. John Lateran. Monte Cassino lay in ruins 140 years, during which time the great Order had its chief seat in Rome.

Thus did Rome and Italy learn what they had gained by reunion with the eastern empire under Justinian. The pitiless financial exaction of that empire was exerted wherever it had power. War and pestilence ravaged town and country. It cost the Church a labour of 200 years to turn the Lombards from Arians and savages into Catholics who should one day be capable of resisting a Barbarossa and generating a Dante.

What, during these 200 years, an imperial exarch at Ravenna was like Gregory tells us in a letter to his friend Sebastian, bishop of Sirmium: "Words cannot express what I suffer from your friend, the lord Romanus. I may say that his malice against us is worse than the swords of the Lombards. The enemies who slay us seem to us kinder than the magistrates of the commonwealth, who wear our hearts out with their malignity, their plundering, and their deceit. At one and the same time to superintend bishops and clergy, monasteries also and the people, carefully to watch against insidious attacks of our enemies, and be perpetually on guard against the treachery and ill-treatment of our rulers, you, my brother, can the better judge what labour and sorrow is here in proportion to the purity of your affection for me who suffer it."[1]

This glimpse will be enough of the generation which preceded the accession of St. Gregory to the Chair of Peter. The whole fifty years of his life up to that time were for his country like the prophet's scroll, inscribed with lamentation and mourning and woe. And in his words to the bishop of Sirmium he gives a faithful picture of the position which his successors held until the time when at length they invoked the king of the Franks to come to the succour of St. Peter.

The calamities which fell upon Italy, and especially upon Rome, in the five captures of the Gothic war, in the subsequent descent of the Lombards, in the subjection of the old capital to a distant and despotic lord, were so great that eye-witnesses declare no language could express them. That they were to the Popes

[1] *Ep.* v. 42, p. 769.

themselves unspeakably distressing, that the Popes did all in their power to avert them, the letters of the Popes remain to testify. I must now dwell for a time on the singular result which they had upon the Roman Primacy. When temporal calamities less than these fell upon the cities of Alexandria and Antioch, the seats of the other two original Petrine patriarchates, the authority of their prelates sunk almost to nothing. Before these calamities they had yielded up a large portion of their dignity and autonomy to the over-reaching see of the eastern capital, the rank of which, above that of a simple bishopric, rested on nothing but the emperor's will to concentrate spiritual power in his own hands, by making its seat for the whole eastern empire the city of the Bosporus. But when Rome was ruined in the Gothic war nothing of the kind took place. St. Gregory inherited his place as successor of St. Peter without the least impairment of the authority which his see had held from the beginning. One wound, indeed, had been inflicted upon it by the Herule Odoacer, when in occupation of the sovereign power which he held over Italy, in name, by delegation of the emperor Zeno, in fact, as head of the foreign mercenaries, he had claimed a right to confirm the election of the Pope when chosen. Theodorick and Theodatus had continued to exert that right—and from the Goths Justinian had taken it—and Gregory himself, as we have seen, had applied to the imperial power at Constantinople to frustrate his own election by clergy and people. But the Pope, when once recognised, entered upon his full and undiminished authority. All that St. Leo had been St. Gregory was, though Rome had been almost destroyed, and was in the temporal rule subject to the emperor's officer, the exarch at Ravenna. I do not know any fact of history which brings out more distinctly the character of the Pope as inheriting the charge over the whole Church committed by our Lord to St. Peter. That was not a charge depending on the city in which it might be exercised. It was a charge committed to the chief of the Apostles. As our Lord promised to be with the apostolic body to the consummation of the world, as all their spiritual powers depended on His being with them, so, above all, most of all, the spiritual power of their head. Rome might be absolutely destitute of inhabitants after Totila's victory, but the Pope was not touched. Rome might cease to be capital even of a province, but the Pope was not touched. And it was a series of the most terrible disasters which revealed this prerogative of the Pope as head of the Christian hierarchy. The Pope might be a captive at Constantinople, scorned, deceived, torn away even from the refuge of the altar, surrounded with spies, betrayed by subservient bishops and patriarchs, and, worst of all, be labouring under the stigma of an election originally enforced by arbitrary violence; a despotic emperor might do his worst, but the Pope's successors carried on his prerogatives unimpaired. The walls of Aurelian preserved Rome from the Lombard, but the Pontiff who kept guard over them was not contained in them.

His rule was intangible by material attack as it was beyond the reach of material despotism. Italy might be ruined, and a new Rome made out of its ruins, but the Pope would be the maker of it. And the most terrible calamity was chosen to reveal this singular prerogative. The death of *Senatus populusque Romanus* discovered even to the outside world the life which proceeded from St. Peter's body, as each archbishop received from St. Peter's successor the pallium which had been laid upon it. Thus was conveyed to the mind by the senses that participation of the Primacy, in which consisted all the authority which he exercised over other bishops. The violence of the Teuton, the misbelief of the Arian, the despotism of the Byzantine, were unconsciously co-operating to this result.

For it must be added that the Rome which survived after the conquest by Justinian only lived by the Primacy of which it was the seat. Two historians[1] of the city, writing from quite opposite points of view, one a Catholic Christian, the other a rationalistic unbeliever, unite in witnessing that from the time of Narses the spiritual power of the Primacy was the spring of all action. Not only such new buildings as arose were churches and the work of the Popes; St. Gregory also fed the city from the patrimonium of the church which he administered. Rome had been made by her empire, which the political wisdom and valour of her citizens had formed through so many centuries. When at length the wandering of the nations had broken up

---

[1] Reumont and Gregorovius.

that empire, and the northern soldiers whom the emperors, specially from Constantine onwards, had enrolled in her armies and taken for their ministers and generals, followed the example of Alaric and Ataulph, and assumed the rule for themselves, the situation of Rome offered it no protection. The emperor who, at the beginning of the fifth century, took refuge from Alaric in Ravenna was followed a century later by the Gothic king, whose body, still reposing in his splendid tomb at Ravenna, was a memorial that this fortress had been the centre of his power. Theodorick was succeeded by the exarch, the permanent representative of an absent lord. We are following the fortunes of Rome in the 300 years from Genseric to Astolphus. In the second and third of these three centuries Rome would have ceased to exist, but for the imperishable life which did not come from her but was stored up in her. That life was the *form* of her new body; otherwise it would have been a carcase lying prostrate in the dust of mouldering theatres and desolated baths. Their patriarchs saved neither Antioch nor Alexandria; but the Papacy not only saved Rome, but created her anew.

Out of such a Rome St. Gregory poured forth his sorrows to the empress Constantine, wife of Mauritius: "It is now seven-and-twenty years since we have been living in this city among the swords of the Lombards".[1] He was writing in the year 595, and he reckons from the descent of Alboin in 568. "What the sums called

[1] *Ep.* v. 21, p. 751.

for from the Church in these years day by day to live at all have been I cannot express. I may say in a word that as your Majesties have, with the first army of Italy at Ravenna, a chancellor of the exchequer who supplies daily wants, so in this city for the like purpose I am such a person. And yet this same church which at one and the same time is at such endless expense for the clergy, the monasteries, the poor, the people, and moreover for the Lombards, is pressed also by the affliction of all the churches, which groan over the pride of this one man, yet do not venture to utter a word."

And Gregory, referring just before to the pride of this one man, who had the audacity to put in a letter to the Pope himself, a superscription in which, according to the Pope's judgment, he claimed to be sole bishop in the Church, used words which will serve to indicate what Gregory conceived his own authority to be, as well as the source on which it rested: "I beseech you, by Almighty God, not to permit your Majesty's time to be polluted by one man's arrogance. Do not in any way give your consent to so perverse an appellation. By no means let your Majesty in such a cause despise me the individual, for the sins of Gregory are indeed so great as to deserve such treatment, but there are no sins of the Apostle Peter that he should deserve in your time such treatment. Wherefore, I again and again entreat you, by Almighty God, that as former princes, your progenitors, have sought the favour of the holy Apostle Peter, so you also would seek it and preserve it for yourselves. Nor let his honour be in your mind the least diminished by our sins, his unworthy servant: that he may be now your helper in all things, and hereafter be able to pardon your sins."

I quote the following passage from a letter[1] to the emperor Mauritius himself, not only because Gregory alleges as the root of his own authority the three great words spoken by our Lord to Peter, but for the description of the times in which he lived, and the vast importance of union between the two great powers. This, he says, if faithfully maintained on both sides, would have protected them from such calamities.

"Your Majesty, who is appointed by God, watches, among the other cares of your empire, with the uprightness of a spiritual zeal over the preservation of sacerdotal charity. For, with piety as well as truth, you think that no one can rule well the things of this world unless he knows how to treat divine things, and that the peace of the human commonwealth depends on the peace of the universal Church. For, most gracious emperor, what power of man, what masterful arm of flesh, would presume to lay unholy hands upon the dignity of your most Christian empire, if the bishops were with one accord of mind to beseech their Redeemer for you by their words, and, if need be, by their deservings? Is there any nation so ferocious as to use its sword so cruelly for the destruction of the faithful, unless our life, who are called but are not bishops, had

[1] *Ep.* v. 20, tom. ii. 747.

upon it the stain of the worst actions? While, deserting what belongs to us, and aiming at what is beyond us, we add our own sins to the brute strength of barbarians. Our guilt sharpens the swords of our enemies, and weighs down the strength of the State. What excuse can we make who press down the people of God, over which we unworthily preside, with the burden of our sins? Who preach with our tongues and kill by our examples? Whose works teach iniquity, while their words make a show of justice? We wear down the body with fasts, while the mind swells with arrogance. This puts on poor apparel; that has more than imperial pride. We lie in ashes, and despise dignities. We teach the humble, and lead the proud, and hide the wolf's teeth in the sheep's face. What result has all this but that, while we impose on men, we are made known to God? Thus it is with the greatest wisdom that your Majesty seeks the peace of the Church as the means of stilling the tumults of war, and would make the hearts of bishops rest once more in its solid structure. That is my wish: in that to the utmost of my power I obey you.

"But since it is not my cause but God's, and since not I only but the whole Church is thrown into confusion; since sacred laws, since venerable councils, since the very commands even of our Lord Jesus Christ are disturbed by the invention of this haughty and pompous language, let the most pious emperor lance the wound and overcome the sick man's resistance by the force of the imperial authority. If you bind up that wound, you raise up the State; and by cutting off such abuses, contribute to the length of your reign.

"For to all who know the Gospel it is notorious that the charge of the whole Church was entrusted by the voice of the Lord to the holy Apostle Peter, chief of all the Apostles." And he then cites, as so many of his predecessors cited, the three great words. He concludes: "Peter received the keys of the kingdom of heaven, the power of binding and loosing, the charge of the whole Church, the Principate over it; yet he is not called the universal Apostle, and John, my colleague as bishop, endeavours to be called universal bishop.

"All things in Europe are delivered over to the power of barbarians. Our cities are destroyed, our fortresses overthrown, our provinces depopulated. The ground remains untilled. Day by day idolaters exercise their rage upon the faithful, who are cruelly slaughtered; and bishops who should lie in dust and ashes seek for themselves vanitous names: glory in new and profane titles.

"Am I in this defending a cause proper to myself? Am I resisting my own special injury? Nay, it is the cause of Almighty God: the cause of the universal Church. Who is he who, in spite of the commands of the Gospel, in spite of the decrees of councils, presumes to usurp a new title for himself? I would that he who has agreed to be called universal may be himself one, without the diminution of others.

"And we know, indeed, that many bishops of Constantinople have fallen into the gulf of heresy; have become not heretics only but heresiarchs. Thence came Nestorius, who, deeming Jesus Christ, the Mediator of God and man, to be two persons, because he did not believe that God could become man, went even to the extent of Jewish unbelief. Thence came Macedonius, who denied the Godhead of the Holy Spirit, consubstantial with the Father and the Son. If, then, anyone seizes upon that name for himself, as in the judgment of all good men he has done, the whole Church—which God forbid—falls from its state when he who is called universal falls. But far from the hearts of Christians be that blasphemous name in which the honour due to all bishops is taken away, while one madly arrogates it to himself.

"I know that in honour of St. Peter, prince of the Apostles, that title was offered to the Roman Pontiff during the venerable Council of Chalcedon. But no one of them ever consented to use this name of singularity: lest while something peculiar was given to one, all bishops should be deprived of the honour due to them. Do we, then, not seek the glory of this name, even when offered to us, and does another catch at it for himself, when it is not offered?

"Your Majesty, then, must bend that neck which refuses obedience to the canons. He must be restrained who does an injury to the whole Church; who is proud in heart; who has a greed after a name given to none other; who by such a singular name throws a slur upon your empire also in putting himself over it.

"We are all scandalised at this: let the author of

the scandal return to right, and all contest between bishops will cease. For I am the servant of all bishops so long as they live like bishops. But whoever, through vainglory and contrary to the statutes of the Fathers, lifts his neck against Almighty God, I trust in Almighty God that he will not bend me even with the sword."

As Gregory quotes the three words said to Peter, with application of them to his own see, it seems needless to repeat other passages in which he says the same thing. But there is a letter to Eulogius,[1] patriarch of Alexandria, which begins by saying that this patriarch had written to him much concerning the See of Peter, and that he sat in it in his successors down to Gregory's own time. Whereupon Gregory, before himself citing the three words, says: "Who does not know that holy Church is founded on the solidity of the chief Apostle, whose name expressed his firmness, being called Peter from Petra". Then he calls the attention of Eulogius to the fact that all the three patriarchal sees were sees of Peter, with this remarkable inference, that "though there were many Apostles, only the see of the prince of the Apostles, which is the see of one in three places, received supreme authority *in virtue of its very principate*".[2]

Let us attempt to gather the meaning of the various statements quoted from St. Gregory, and see whether they do not form a coherent whole.

---

[1] *Ep.* vii. 40, p. 887.

[2] I have drawn attention to this fact, and the idea which it represents as attested by Popes earlier than St. Gregory, in vol. v., pp. 53-60, of the *Formation of Christendom*, "The Throne," &c.

He claims, like all his predecessors, the three great texts concerning Peter, as conveying the charge of the whole Church, the Principate, to Peter and his heirs, that is, the Popes preceding him.

He contrasts in the most pointed manner this charge with the name of Ecumenical, which he translates universal, patriarch, as assumed by the bishop of Constantinople, and he contrasts not the name only, but the thing which he conceives to be meant by the name and carried in it.

He contrasts likewise the moderation of his predecessors, who, though inheriting Peter's charge over the whole Church, declined to accept a name which seemed to exclude other bishops from their proper honour.

Peter's charge over the whole Church, then, in the judgment of Gregory, had descended to himself, as he wrote to the empress, "though the sins of Gregory, who is Peter's unworthy servant, are great, the sins of the Apostle are none," to justify the treatment he has met with in this assumption by another of the title Ecumenical. In a word, the *charge* is a command of the Gospel, the *assumption* is "a name of blasphemy and diabolical pride, and a forerunner of Antichrist".

I conceive that we may interpret St. Gregory's mind in this way. When he so wrote he had behind him rather more than five full centuries since St. Peter and St. Paul had given up their lives in Rome for the Christian faith, and become its patron saints. In all that time Gregory had seen the hierarchy founded by the bearer of the keys fill the earth. Peter, as a token of his Principate, had put his name in the three chief sees, sitting himself as bishop in Antioch for seven years; sitting also himself in Rome, as bishop, and dying there; sending also his disciple Mark from Rome to Alexandria. Our Lord's gift and charge to Peter was the source of unity in His Church. He Himself being mediator between God and man united His Church with the Divine Trinity in unity. Then He gave the keys of His kingdom to Peter, in whom unity was secured through the three patriarchs and the other bishops. Such was the constitution which stood without a break before St. Gregory from the Apostles to the Nicene Council. From St. Sylvester to his own time the Popes had been maintaining that constitution. But now the claim of the bishops of Constantinople was directly against this constitution. Pope Gelasius, his predecessor, had told that bishop in his day that he had no rank above that of a simple bishop.[1] For all their adventitious rank they rested, not upon God, not upon Jesus Christ, not upon St. Peter, but upon the residence of the emperors in their city. That was the ground upon which they called themselves ecumenical, a title which Gregory interpreted universal. Their first step in moving beyond the position of simple bishop was when the 150 bishops at Constantinople in 381 attempted to give them the second place in rank. And this they did not upon any ground of apostolic descent, but because Constantinople

---

[1] Rump, ix. 501-2 ; see his words quoted above, p. 107.

was Nova Roma. As to their act in doing this Gregory writes to Eulogius: "The Roman Church up to this time does not possess, nor has received, the canons or the acts of that council; it has received that council so far as it condemned Macedonius".[1] Their next step was at the Council of Chalcedon to attempt passing a canon, to the effect that the Fathers had given its rank to Rome because it was the capital, that the 150 Fathers had therefore given the second rank to Constantinople, because it was the *new* capital; and that, therefore, the Pontic, the Arian, and the Thracian exarchs of Cæsarea, Ephesus, and Heraclea should be subjected to it. This canon St. Leo had absolutely rejected, and the emperor Marcian had accepted his rejection. In the 130 years from St. Leo to himself, St. Gregory had seen the assumptions of the bishops of Constantinople continually increasing. They rested upon the imperial favour. And now in the case of John the Faster they had gone so far that he prefixed his assumed title of ecumenical patriarch to the very documents which he sent to the Pope for revision. And this though the cause had been settled by himself, and had now come before the Pope, whose power therefore to revise the sentence of one who called himself ecumenical patriarch he did not dispute.

Nor, indeed, did it appear over what domain he claimed to be universal. It might be over the eastern bishops; it might be over the two patriarchs of Alexandria and Antioch, with the later patriarch of

---

[1] *Ep.* vii. 34, p. 882.

Jerusalem; it might be over the actual Roman empire; it might be, finally, over the whole Church. But whichever it might be, the claim would equally be, in Gregory's judgment, unlawful, based simply and solely upon imperial power; resting also in its origin upon a direct untruth, which assaulted the whole foundation whereon the charge of the whole Church, the Principate of Gregory, rested; couched, moreover, in language which would enable future generations of Greeks to draw the conclusion that, since the Primacy of Rome proceeded from its being the capital, when Rome ceased to be the capital, and Constantine's city became the capital, the Primacy also passed to it.

Thus, in the whole assumption of the bishops of Constantinople, it was presupposed that the spiritual power and the hierarchy of the Church descended not from Jesus Christ, but from the emperors.[1] So it is clear that this empty title, which seemed to the emperor Mauritius a meaningless word, a mere nothing, contained in itself the whole system of Antichrist. The Pope saw it, and his words are the more significant when we remember that at the time he uttered them the man had already reached full manhood who was to cut the empire of Justinian in half, to deprive of their liberty three of the eastern patriarchs, destroy a multitude of the Christian people, and be parent of the religion which through the course of 1200 years has shown itself to be specially anti-Christian. There in his Arab tent, as yet the faithful husband of an old wife,

[1] Rump, ix. 502.

was the future Khalif, in whom the spiritual and the temporal power would be joined together; who would set up in a false theocracy that usurpation which Constantine's eastern successors were striving to carry out in the Christian Church. Mahommed would consecrate that very false principle which was at the root of the ecumenical patriarch's arrogance. Thus the strongest word used by Gregory of John the Faster's assumption, that it was "a name of blasphemy, of diabolical pride, and a forerunner of Antichrist," received its exact verification within a generation after Gregory had spoken it.

But Gregory's charge and Principate were of divine creation, and did not exclude the proper power and jurisdiction either of every bishop or of the whole episcopate, at the head of which it stood, and through which it worked, carefully maintaining what had been from the beginning, preserving the rank and place of each, consolidating all in the one structure.[1] The intruder set up by the imperial power deposed Alexandria and Antioch to make them subject to himself; the lawful shepherd maintained Alexandria and Antioch because they grew upon the tree of which he was the trunk. His charge did not exclude, but did indeed include them. The reasoning of St. Gregory in his letter to the emperor of the day, and his very words in his letter to the patriarch Eulogius, have become a

---

[1] Providentissime piissimus Dominus ad compescendos bellicos motus pacem quærit ecclesiæ *atque ad hujus compagem sacerdotum dignatur corda reducere.* —*Ep.* v. 20, p. 747.

matter of faith by their enrolment in the decree of the Vatican Council. That decree defines the Principate to be an episcopal power of jurisdiction, which is immediate, over the whole Church. By it the whole Church becomes one flock, under one shepherd. And it further defines that, "It is so far from being true that this power of the Supreme Pontiff is injurious to the ordinary and immediate power of episcopal jurisdiction, by which bishops placed by the Holy Spirit have succeeded the Apostles, and as true pastors feed and rule the flocks severally assigned to them, each his own, that this jurisdiction is asserted, strengthened, and maintained by the supreme and universal pastor, according to St. Gregory's words : 'My honour is the honour of the universal Church ; my honour is the solid strength of my brethren ; then am I truly honoured when his due rank is given to each'."[1]

It may be observed that Gregory's position against the assumption of John the Faster is the same as St. Leo's position against Anatolius. In both cases the Popes discerned the hostile power located in the see of Nova Roma which was at work against the original order of the Church, and the Pope who was at the head of it. The only difference lies in the great advance which the hostile power had made on one hand, and on the other hand the excessively difficult temporal position in which St. Gregory had to fight the battle for the cause, as he said, of the universal Church. Yet

[1] De vi et ratione Primatus Romani Pontificis—c. iii., quoting the letter of St. Gregory to Eulogius, viii. 30.

the speech of the Pope beleaguered by the Lombards in a decimated and subject Rome is as strong as the speech of the Pope who had the imperial grandchildren of Theodosius for friends and supporters, and, when they failed, saved Rome by her two Apostles from the destruction menaced by Attila and Genseric.

But there was no one in the eastern Church—neither the emperor Mauritius, nor the patriarch John the Faster, nor the patriarch Eulogius—who failed to acknowledge the Pope's charge over the whole Church, grounded on the three texts to Peter. Gregory himself reprehends the patriarch Eulogius for giving him in the superscription of his letter the title "universal Pope". He chose for himself, in opposition to the bishop John's arrogated title of ecumenical patriarch, that of "servant of the servants of God". The title chosen indicated the temper in which St. Gregory exercised the vast charge which he had inherited. For if there is any one principle which seems to serve as the favourite maxim of his whole pontificate, it is that expressed in a letter to the bishop of Syracuse. That bishop had been speaking of an African primate who had professed that he was subject to the Apostolic See. St. Gregory's comment is: "If a bishop is in any fault, I know not any bishop who is not subject to it. But when no fault requires it, all are equal according to the estimation of humility."[1] Natalis, archbishop of Salona, in Dalmatia, had given the Pope much trouble. The Pope deals with him tenderly in more than one

---

[1] *Ep.* ix. 59, p. 976.

letter. But he says: "After the letters of my predecessor (Pelagius) and my own, in the matter of Honoratus the archdeacon, were sent to your Holiness, in despite of the sentence of us both, the above-mentioned Honoratus was deprived of his rank. Had either of the four patriarchs done this, so great an act of contumacy could not have been passed over without the most grievous scandal. However, as your brotherhood has since returned to your duty, I take notice neither of the injury done to me, nor of that to my predecessor."[1]

Of the immense energy shown by St. Gregory in the exercise of his Principate, of the immense influence wielded by him both in the East and in the West, of the acknowledgment of his Principate by the answers which emperor and patriarch made to his demands and rebukes, we possess an imperishable record in the fourteen books of his letters which have been preserved to us. They are somewhat more than 850 in number. They range over every subject, and are addressed to every sort of person. If he rebukes the ambition of a patriarch, and complains of an emperor's unjust law, he cares also that the tenants on the vast estates of the Church which his officers superintend at a distance should not be in any way harshly treated. He writes to his *defensor* in Sicily: "I am informed that if anyone has a charge against any clerks, you throw a slight upon the bishops by causing these clerks to appear in your own court. If this be so, we expressly order you to

---

[1] *Ep.* ii. 52, p. 618.

presume to do so no more, because beyond doubt it is very unseemly. If anyone charges a clerk, let him go to his bishop, for the bishop himself to hear the case, or depute judges. If it come to arbitration, let the so-deputed judges cause the parties to select a judge. If a clerk or a layman have anything against a bishop, you should act between them either by hearing the cause yourself, or by inducing the parties to choose judges. For if his own jurisdiction is not preserved to each bishop, what else results but that the order of the Church is thrown into confusion by us, the very persons who are charged with its maintenance.

"We have also been informed that certain clerks, put into penance for faults they had committed by our most reverend brother the bishop John, have been dismissed by your authority without his knowledge. If this is true, know that you have committed an altogether improper act, worthy of great censure. Restore, therefore, at once those clerks to their own bishop, nor ever do this again, or you will incur from us severe punishment."[1]

I have quoted already his letters on eastern affairs. They might be enlarged upon to any extent. As to those who held the highest rank, he has warm sympathy with a deposed patriarch of Antioch, sending him a copy of the letter which announced his accession, as well as to the sitting patriarchs. After twenty years' deposition Anastasius was restored. He has also close friendship with Eulogius, patriarch of Alexandria, to

[1] *Ep.* xi. 37, p. 1120.

whom he writes gracefully : " Besides our mutual affection, there is a peculiar bond uniting us to the Alexandrian Church. All know that the Evangelist Mark was sent by his master Peter; thus we are clasped together by the unity of the master and the disciple. I seem to sit in the disciple's see for the master's sake, and you in the master's see for the sake of the disciple. To this we must add your personal merits; for we know how you follow the institutions of him from whom you spring. Thus we are touched with compassion for what you suffer; but we shrink from telling you what we endure ourselves by the daily plundering, killing, and maiming of our people by the Lombards."[1]

Let us here take a short view of Gregory's incessant activity among the western nations in process of formation. In his struggle to tame the ferocity, lawlessness, and unbelief of the Lombards, he betakes himself to the illustrious Catholic queen Theodelinda. He strives to use her influence with her husband Agilulf, on behalf of Rome, ever the object of oppression. Knowing her to be a good Christian, he sent her his *Dialogues*. He also set before her the supremacy of his see, because she had been misled into withdrawing from the communion of the new archbishop of Milan, Constantius. The Pope assures her that the archbishop, as well as himself, venerates the doctrinal decisions of the Four Councils. He adds: " Since, then, by my own public profession you know the entireness of our belief, it is fitting that you have no further scruple concerning the

[1] *Ep.* vi. 60, p. 836.

Church of St. Peter, prince of the Apostles. But persist in the true faith, and ground your life on the rock of the Church, that is, in his confession: lest your many tears and your good works avail nothing, if they be separated from the true faith. For as branches wither without a root, so works, however good they seem, are nothing if separated from the solidity of the faith."[1]

Ten of his letters are addressed to Brunechild, the terrible queen of the Franks. But his letter to all the Gallic bishops in the kingdom of Childebert will best set forth his authority. That king then reigned over nearly all France. The Pope began by saying that the universe itself was ruled by graduated orders of spirits. If there was such distinction of ranks even in the sinless, what man should hesitate to obey a disposition to which angels are subject? "Since, then, each individual office is happily fulfilled when there is a superior to whom application can be made, we have thought it good, following ancient custom, to make our brother Virgilius, bishop of Arles, our representative in the churches which are in the kingdom of our most illustrious son king Childebert. We do this in order that the integrity of the Catholic faith, that is, of the Four holy Councils, may by God's protection be carefully preserved; and that, if any contention should arise between our brethren and fellow-bishops, he may, by virtue of his authority, as holding the place of the Apostolic See, reduce it by discreet moderation. We

---

[1] *Ep.* iv. 38, p. 718.

have also enjoined him, that if any contest should arise requiring the presence of others, he should collect a sufficient number of our brethren and fellow-bishops, discuss the matter equitably, and determine it in conformity with the canons. But if, which the divine power avert, contest should arise on a matter of faith, or some business emerge about which there is great hesitation, and which for its magnitude requires the judgment of the Apostolic See, after diligent examination of the facts, he is to make report to us, that we may terminate all doubt thereon by a fitting sentence."[1]

In this letter we are at a hundred years after the conversion of Clovis. The Catholic kingdom has swallowed up its Arian competitors whether at Toulouse or at Lyons, and over it stands the protecting vigour of Gregory, as a hundred and fifty years before that of Leo strove to support the falling empire. Arles receives the pallium for the Frankish kingdom, as it held it for the Theodocian empire, from Rome. Leo saw the imperial line expire at Rome; from Rome Gregory places the bishops "of his most illustrious son Childebert" under the old primacy of Arles. This is the "solidity" of the rock of Peter in which Gregory recommends the queens Theodelinda and Brunechild to place themselves.

We know how Gregory, while yet a Roman deacon and monk, walking one day from the palace which he had made a monastery, scarcely more than a stone's-throw to the forum in which a slave-market was held,

[1] *Ep.* v. 54, p. 784.

was moved to pity at the sight of the fair-haired Angles; how he was minded to leave Rome himself on a mission to convert them; how he was kept back by the affection of the Romans; how Pope Pelagius suddenly died of the plague, and Gregory, in spite of all his efforts, was made to succeed him; how from the See of Peter he sent out Augustine and his forty monks to the lost island in the Atlantic, where, since Stilicho withdrew the Roman armies, every cruelty had revelled, and every pagan abomination had been practised by the Saxon invaders. To many, no doubt, the subsequent success of Gregory's venture to convert the Anglo-Saxon England has served to disguise its danger and difficulty at the time. When Augustine reached the shores of Kent, the successive invasions of the Saxon pirates had set up eight petty kingdoms upon the ruin of the Roman civilisation and the Christian Church. The miseries which are covered under those five generations of unrecorded strife are supposed to have exceeded the misery endured in France, Spain, Italy, and the Illyrian provinces during the same time. The old inhabitants were reduced to slavery, or exterminated, or driven to the three corners of Cornwall, Wales, and Strathclyde. So bitter was the British feeling under the destruction of their country and the wrongs they had endured, that it overcame all Christian principle in them, and the Welsh refused all aid to the Roman missionary in the attempt to convert a race so cruel. It required all St. Gregory's firmness to induce his own monks to persist. In all the annals of Christian enterprise during eighteen centuries, there is probably not one which presented less hope of success than St. Gregory's resolution to add the spiritual beauty of the Christian to the physical beauty which he admired in the captives of the Roman forum.

Among those to whom he applied to assist and further his purpose was the great queen of the Franks. To Brunechild he directed a letter saluting her, he says, with the charity of a father: "We hear that, by the help of God, the English people is willing to become Christian; and we recommend the bearer of these, the servant of God, Augustine, to your Excellency, to help him in all things, and to protect his work".[1]

It was also to Virgilius, bishop of Arles, and primate of all the Gallic bishops, as we have seen, by Gregory's own appointment, that he sent Augustine, after his first success with Ethelbert, to receive episcopal consecration.

From Gregory's own hand, and in virtue of his apostolic power, England in its second spring received its division into two provinces, one to be seated at Canterbury, the other at York. His letters to St. Augustine still exist to show how he entered into all the difficulties of the missionary, all the needs of a land in conversion from paganism. From him date the great prerogatives of the see of Canterbury, extending over the whole island, inasmuch as it was the matrix of the Church in England. If sons may deny their father, Englishmen may deny Gregory, and add to schism the guilt of parricide.

[1] *Ep.* vi. 59, p. 835.

But Gregory was hardly less active in restoring Spain from the Arian blight than in giving birth to a new Christian England. He writes, in 594: "We have heard from many who have come from Spain how lately Hermenegild, son of Leovigild, king of the Visigoths, has been converted from the Arian heresy to the Catholic faith by the preaching of Leander, bishop of Seville, long united to me in intimate friendship. His Arian father, by bribes and threats alike, tried to bring him back. Not succeeding, he deprived him of his rank and all his possessions. When this also failed, he put him in close imprisonment, fettering both neck and hands. So Hermenegild learnt to despise the earthly kingdom, and to yearn after the heavenly, while he lay in bonds and sackcloth. When Easter came, his father sent him in the middle of the night an Arian bishop that he might receive communion sacrilegiously consecrated, and so recover his favours. Hermenegild repulsed the bishop with strong reproaches. The father, hearing his report, burst into fury and sent officers to destroy him. They split open his skull with an axe, and so destroyed the life of the body which he had disregarded. Miracles followed. Psalms were heard about the body of the royal martyr—royal, indeed, because he was a martyr."[1]

Writing to St. Leander, archbishop of Seville, Gregory says: "I am so tossed by this world's waves that I cannot steer to harbour this old weather-beaten bark which the secret dispensation of God has com-

[1] *Dialog.*, iii. 31, p. 345, A.D. 594.

mitted to my care. Shipwreck creaks in its worn-out planks. Dearest brother, if you love me, stretch out the hand of your prayers to me in this tempest. Your reward for helping me will be greater success in your own labours.

"No words of mine can express the joy which I feel at hearing the perfect conversion of our common son, king Rechared, to the Catholic faith."[1]

On another occasion Gregory writes to Leander, sending him the pallium, "blessed by Peter, prince of the Apostles," only to be used at Mass: "I see by your letter that burning charity which kindles others. He who is not himself on fire cannot inflame others. I always call to mind your life with great veneration. But as for me I am not what I was: 'Call me not Noemi, which is fair; call me Mara, for I am full of bitterness'. Following the way of my Head, I had resolved to be the scorn of men, the outcast of the people. But the burden of this honour weighs me down; innumerable cares pierce me like swords. There is no rest of the heart. I was tranquil in my monastery. The tempest arose; I am in its waves, suffering with the loss of quiet a shipwreck of mind. The gout oppresses you; I also am terribly pained by it. It will be well if, under these strokes of the scourge, we perceive them to be gifts, by which the sense of the flesh may atone for sins which delights of the flesh may have led us to commit.

"The shortness of my letter will show how weak and

[1] *Ep.* i. 43, p. 531.

how occupied I am, who say so little to one whom I love so much."¹

St. Gregory tells us that king Rechared, after the martyrdom of his brother St. Hermenegild, was converted from the Arian heresy, and brought the whole Visigothic nation to the Catholic faith. "The brother of a martyr fitly became a preacher of the faith. If Hermenegild had not died a martyr, this he would not have been able to do; for 'except the grain of wheat falling into the ground dieth, itself remaineth alone; but if it die, it bringeth forth much fruit'. This we see to be doing in the members which we know to have been done in the Head. In the nation of the Visigoths one died that many might live."²

A letter of St. Gregory to this king Rechared is extant, which one of the greatest French bishops, Hincmar of Reims, nearly three hundred years after it was written, thought worthy to be sent as a present to the emperor Charles the Bald. I quote portions of it:³

"Most excellent son, words cannot tell the delight which I receive from your work and from your life. When I hear the power of that new miracle wrought in our days, that by means of your Excellency the whole nation of the Goths has been brought over from the error of the Arian heresy to the solidity of the right faith, I exclaim with the prophet, 'This is the change of the hand of the Most High'. Is there a heart of stone which would not be softened on hearing of so

---

[1] *Ep.* ix. 121, pp. 1026-8, shortened.
[2] *Dialog.*, iii. 31, p. 348.   [3] *Ep.* ix. 122, p. 1028.

great a work into praises of Almighty God and affection for your Excellency? Often, when my sons meet, it is my pleasure to tell them of the deeds wrought by you, and to join my admiration with theirs. I get angry with myself that I am lazy, useless, and inert, while kings are labouring for the gain of the heavenly country by the ingathering of souls. What, then, shall I allege to the Judge at that tremendous tribunal, if I come before Him then with empty hands, while your Excellency leads a long train of the faithful whom you have drawn into the grace of the true faith by zealous and continuous preaching? But by God's gift this is my great consolation, to love in you that holy work which I have not in myself. When your acts move me to a great exultation, I make mine by charity what is yours by labour. Thus, in your work and our exultation over it, we may cry out with the angels over the conversion of the Goths, 'Glory to God in the highest, and on earth peace to men of good will'. But how joyfully St. Peter, prince of the Apostles, has received your offerings is borne witness to all men by your life.

"You tell me that the abbots, who were carrying your offering to St. Peter, were driven back by a bad sea passage into Spain. Your gifts, which afterwards arrived, were not refused, but the courage of their bearers was tried. The adversity which good intentions encounter is a trial of virtue, not a judgment of reprobation. When St. Paul came to preach in Italy, how great was the blessing he brought; yet he was shipwrecked in coming, but the ship of his heart was not broken by the waves of the sea.

"Also, I am told that your Excellency issued a certain decree against the misbelief of the Jews, which they strove by a bribe to have modified. This bribe you despised, and in the desire to please God preferred innocence to gold. This brought to my mind king David's act. He longed for a draught from the fountain of Bethlehem, which the enemy's host encompassed. His soldiers risked their lives to bring it. But he refused, saying: 'God forbid that I should drink the blood of these men. So he offered it to the Lord.'[1] If an armed king made a sacrifice to God of the water which he refused, think what a sacrifice to Almighty God that king presented who for His love refused to receive, not water, but gold. Therefore, most excellent son, I say confidently that the gold which you refused to receive against God you offered to Him. These are great deeds, the glory of which is due to God. . . .

"Government of subjects should be tempered with great moderation, lest power steal away the judgment. A kingdom is ruled well when the glory of ruling does not overmaster the spirit. Provide also against fits of anger, lest unlimited power be used hurriedly. Anger in punishing even delinquents should not anticipate judgment like a mistress, but follow reason as a servant, coming when she is called. If it once is in possession of the mind, it puts down to justice even a cruel deed. Therefore it is written: 'The

[1] Paralipom. i. 11, 18.

wrath of man worketh not the justice of God'; and again: 'Let everyone be swift to hear but slow to speak'. I do not doubt but that by God's help you practise all this. But as opportunity offers, I creep behind your good works, that when an adviser adds himself to what you do without advice, you may not be alone in your doing. May Almighty God stretch forth His heavenly hand to protect you in all your acts, granting you prosperity in the present life, and, after long years, eternal joy.

"I enclose a small key from the most sacred body of the Apostle St. Peter, with his blessing. It contains an iron filing from his chains, that what bound his neck for martyrdom may deliver yours from all sin. I have also given the bearer of these a cross for you: it contains some of the wood of the Lord's cross, and hair of St. John Baptist; by which you may always be consoled by our Saviour through the intercession of His precursor. To our most reverend brother and fellow-bishop Leander we have sent the pallium from the See of the Apostle St. Peter, in accordance with ancient custom, with your life, with his own goodness and dignity."

This letter of St. Gregory had been drawn forth by one from king Rechared to him, in which the king said he had been minded to inform of his conversion one who was superior to all other bishops, that he had sent a golden jewelled chalice which he hoped might be found worthy of the Apostle who was first in honour. "I beseech your Highness, when you have an opportunity, to find me out with your golden letters. For how truly I love you is not, I think, unknown to one whose breast the Lord inspires, and those who behold you not in the body, yet hear your good report; I commend to your Holiness with the utmost veneration Leander, bishop of Seville, who has been the means of making known to us your good will. I am delighted to hear of your health, and beg of your Christian prudence that you would frequently commend to our common Lord in your prayers the people who, under God, are ruled by us, and have been added to Christ in your times, that true charity towards God may be strengthened by the very distance which divides us."[1]

The fact commemorated in these letters was indeed one for which the Pope might well use the angelical hymn of praise. "The bishops of Spain,"[2] says Gibbon, "respected themselves and were respected by the public; their indissoluble union confirmed their authority; and the regular discipline of the Church introduced peace, order, and stability into the government of the State. From the reign of Rechared, the first Catholic king, to that of Witiza, the immediate predecessor of the unfortunate Roderic, sixteen national councils were successively convened. The six metropolitans—Toledo, Seville, Merida, Braga, Tarragona, and Narbonne—presided according to their respective seniority; the assembly was composed of their suffragan bishops, who appeared in person or by their proxies; and a place was assigned to the most holy or opulent

---

[1] *Ep.* ix. 61, p. 977.
[2] Gibbon, ch. xxxviii. : a sneer or two have been omitted.

of the Spanish abbots. During the first three days of the convocation, as long as they agitated the ecclesiastical questions of doctrine and discipline, the profane laity was excluded from their debates, which were conducted, however, with decent solemnity. But on the morning of the fourth day the doors were thrown open for the entrance of the great officers of the palace, the dukes and counts of the provinces, the judges of the cities, and the Gothic nobles; and the decrees of heaven were ratified by the consent of the people. The same rules were observed in the provincial assemblies, the annual synods which were empowered to hear complaints and to redress grievances; and a legal government was supported by the prevailing influence of the Spanish clergy. . . . The national councils of Toledo, in which the free spirit of the barbarians was tempered and guided by episcopal policy, have established some prudent laws for the common benefit of the king and people. The vacancy of the throne was supplied by the choice of the bishops and palatines; and after the failure of the line of Alaric, the regal dignity was still limited to the pure and noble blood of the Goths. The clergy who anointed their lawful prince always recommended the duty of allegiance; and the spiritual censures were denounced on the heads of the impious subjects who should resist his authority, conspire against his life, or violate by an indecent union the chastity even of his widow. But the monarch himself, when he ascended the throne, was bound by a reciprocal oath to God and his people that he would faithfully execute his important trust. The real or imaginary faults of his administration were subject to the control of a powerful aristocracy ; and the bishops and palatines were guarded by a fundamental privilege that they should not be degraded, imprisoned, tortured, nor punished with death, exile, or confiscation, unless by the free and public judgment of their peers."

We have here the historian, who is one of the bitterest enemies of the Christian Church and Faith, avowing that the barbarian Visigoths received from the hands of that Church and Faith, at the end of the sixth century, the great institutions of a limited Christian monarchy, consecrated by the Church, in which the king at his accession solemnly avowed his responsibility for his exercise of the immense functions entrusted to him; also of parliaments, in which clergy and laity sat together in common deliberation upon the affairs of the State, grievances were redressed, and laws for the benefit of king and people passed; in fact, a reign of legal government, based upon law and justice, and confirmed by religious sanction.

And in all this the hand of the Pope was seen, sending to the chief bishop of Spain the pallium direct from the body of St. Peter, on which it had been laid, as the visible symbol of apostolic power dwelling in the Apostle's See, and radiating from it.

This is the first instance, and not the least striking, of a fact which lies at the foundation of modern Europe ; for so the Teuton war leaders became Christian kings, and so the northern barbarians were changed into Christian nations. For that which Gibbon here describes took place in all the Teuton peoples who accepted the Catholic faith. He has elsewhere said: "The progress of Christianity has been marked by two glorious and decisive victories: over the learned and luxurious citizens of the Roman empire, and over the warlike barbarians of Scythia and Germany, who subverted the empire and embraced the religion of the Romans".[1]

Of this latter victory we can celebrate the accomplishment, as St. Gregory did, in the words of the angelic hymn, but the details have not been preserved for us, even in the scanty proportion which we possess concerning the former. Fighting for thirty years with the Lombards for the very existence of Rome, Gregory was the contemporary and witness of this second victory. Not until the Arian heresy was subdued by the Catholic faith could it be said to be accomplished. The pontificate of his ancestor in the third degree, Pope Felix III., might be called heroic, in that, while under the domination of the Arian Herule, Odoacer, he resisted the meddling with the received doctrine of the Church by the emperor Zeno, guided by the larger mind and treacherous fraud of Acacius, the bishop of Constantinople, who ruled its emperor. Then the Arian Vandals bitterly persecuted the Church in Africa, and the Visigoth Arians had possession of France from the Loire southwards, and of Spain. Nowhere in the whole world was there a Catholic prince. The north

[1] Gibbon, ch. xxxix.

and east of France and Belgium was held by the still pagan Franks. By the time of Gregory, Clovis and his sons had extinguished the Arian Visigoth kingdom and the Arian kingdom of Burgundy, and ruled one Catholic kingdom of all France. Under Rechared, the Arian Visigoth kingdom in Spain became Catholic. Gregory also announced to his friend, the patriarch Eulogius, that the pagan Saxons in England were receiving the Catholic faith by thousands from his missionary. The taint which the wickedness of the eastern emperor Valens had been so mysteriously allowed to communicate to the nascent faith of the Teuton tribes, through the noblest of their family, the Goths, was, during the century which passed between Pope Felix and Pope Gregory, purged away. It was decided beyond recal that the new nations of the West should be Catholic. Five times had Rome been taken and wasted: at one moment, it is said, all its inhabitants had deserted it and fled. The ancient city was extinct: in and out of it rose the Rome of the Popes, which Gregory was feeding and guarding. The eastern emperor, who called himself the Roman prince, in recovering her had destroyed her; but the life that was in her Pontiff was indestructible. The ecumenical patriarch was foiled by the Servant of the servants of God: in proportion as the eastern bishops submitted their original hierarchy, of apostolic institution, and the graduated autonomy which each enjoyed under it, to an imperial minister, termed a patriarch, in Constantinople, all the bishops of the West, placed as they were under distinct kingdoms, found their common centre, adviser, champion, and ruler in the Chair of Peter, fixed in a ruined Rome. If Gregory, in his daily distress, thought that the end of the world was coming, all subsequent ages have felt that in him the world of the future was already founded. In the two centuries since the death of the great Theodosius, the countries which form modern Europe had passed through indescribable disturbance, a misery without end—dislodgement of the old proprietors, a settlement of new inhabitants and rulers. The Christian religion itself had receded for a time far within the limits which it had once reached, as in the north of France, in Germany, and in Britain. The rulers of broad western lands, with the conquering host which they led, had become the victims first, and then the propagators, of the same fatal heresy. The conquered population alone remained Catholic. The conversion of Clovis was the first light which arose in this darkness. And now, a hundred years after that conversion, Paris and Bordeaux, and Toulouse and Lyons, Toledo and Seville, were Catholic once more, and Gregory, a provincial captive in a collapsing Rome, was owned by all these cities as the standard and arbiter of their faith, and the king of the Visigoths thankfully received a few filings from the chains of the Apostle Peter as a present which worthily celebrated his conversion.

It is to be observed that this absolute defeat of the Arian heresy in several countries is accomplished in spite of the power which, in all of them, was wielded by Arian rulers. In vain had Genseric, Hunnerich, Gundamund, and Thrasimund oppressed and tortured the Catholics of Africa, banished their bishops, and set up nominees of their own as Arian bishops in their places for a hundred years. No sooner did Belisarius land on their soil than the fabric reared with every possible deceit and cruelty fell to the ground. The Arian Vandal king was carried away in triumph, as the spoil of a single battle, to Constantinople, and the Catholic bishops, while they hailed Justinian as their deliverer, met in plenary council, acknowledging the Primacy of Peter, as in the days of St. Augustine. In vain had the powerful Visigoth monarchy, seated during three generations at Toulouse, persecuted with fraud and cruelty its Catholic people. A single blow from the arm of Clovis delivered from their rule the whole country from the Loire to the Pyrenees. In vain had Gondeband and his family in Burgundy wavered between the heresy which he professed and the Catholic faith which he admired. The children of Clovis absorbed that kingdom also. But the strongest example of all remains. In vain, too, had Theodorick, after the murder of his rival Odoacer when an invited guest in the banquet of Ravenna, covered over the savage, and governed with wisdom and moderation a Catholic people, whom he soothed by choosing their noblest—Cassiodorus, Symmachus, and Boethius—for his ministers. He had formed into a family compact by marriages the Arian rulers in Africa, Spain, and Gaul. His moderation gave way when he saw the eastern emperor resume the policy of a Catholic sovereign. He put on the savage again, and he ended with the murder not only of his own long-trusted ministers, but of the Pope, who refused to be his instrument in procuring immunity for heresy from a Catholic emperor.

At his death, overclouded with the pangs of remorse, the Arian rule which he had fostered with so much skill showed itself to have no hold upon an Italy to which he had given a great temporal prosperity. The Goths, whom he had seemed to tame, were found incapable of self-government, and every Roman heart welcomed Belisarius and Narses as the restorers of a power which had not ceased to claim their allegiance, even through the turpitudes and betrayals of Zeno and Anastasius.

The best solution which I know for this wonderful result, brought about in so many countries, is contained in a few words of Gibbon: "Under the Roman empire the wealth and jurisdiction of the bishops, their sacred character and perpetual office, their numerous dependents, popular eloquence and provincial assemblies, had rendered them always respectable and sometimes dangerous. Their influence was augmented by the progress of superstition" (by which he means the Catholic faith), "and the establishment of the French monarchy may in some degree be ascribed to the firm alliance of a hundred prelates who reigned in the discontented or independent cities of Gaul."[1] But how were these prelates bound together in a firm alliance? Because each one of them felt what a chief among them, St.

[1] Ch. xxxviii.

Avitus, under an Arian prince, expressed to the Roman senate in the matter of Pope Symmachus by the direction of his brother bishops, that in the person of the Bishop of Rome the principate of the whole Church was touched; that "in the case of other bishops, if there be any lapse, it may be restored; but if the Pope of Rome is endangered, not one bishop but the episcopate itself will seem to be shaken".[1] If the bishops had been all that is above described with the exception of this one thing, the common bond which held them to Rome, how would the ruin of their country, the subversion of existing interests, the confiscation of the land, the imposition of foreign invaders for masters, have acted upon them? It would have split them up into various parties, rivals for favour and the power derived from favour. The bishops of each country would have had national interests controlling their actions. The Teuton invaders were without power of cohesion, without fraternal affection for each other; their ephemeral territories were in a state of perpetual fluctuation. The bishops locally situated in these changing districts would have been themselves divided. In fact, the Arian bishops had no common centre. They were the nominees and partisans of their several sovereigns. They presented no one front, for their negation was no one faith. We cannot be wrong in extending the action assigned by Gibbon to the hundred bishops of Gaul, to the Catholic bishops throughout all the countries in which a poorer Catholic population was

---

[1] See above, p. 141.

governed by Arian rulers. The divine bond of the Primacy, resting upon the faith which it represented, secured in one alliance all the bishops of the West. Nor must we forget that the Throne of Peter acknowledged by those bishops as the source of their common faith, the crown of the episcopate, was likewise regarded by the Arian rulers themselves as the great throne of justice, above the sway of local jealousies and subordinate jurisdictions. It represented to their eyes the fabric of Roman law, the wonderful creation of centuries, which the northern conquerors were utterly unable to emulate, and made them feel how inferior brute force was to civil wisdom and equity.

In the constitution of the Visigothic kingdom of Spain from the time of Rechared, when it became Catholic, we see the first fruits of the Church's beneficent action on the northern invaders. The barbarian monarchy from its original condition of a military command in time of war, directing a raid of the tribe or people upon its enemies, becomes a settled rule, at the head of estates which meet in annual synod, and in which bishops and barons sit side by side. Government reposes on the peaceable union of the Two Powers. In process of time this sort of political order was established everywhere throughout the West, by the same action and influence of the Church. In the Roman empire the supreme power had been in its origin a mandate conferred by the citizens of a free state on one of their number for the preservation of the commonwealth. The notion of dynastic descent was wanting to it from the beginning. But the power which Augustus had received in successive periods of ten years passed to his successors for their life. Still they were rather life-presidents with royal power than kings. And it may be noticed that in that long line no blessing seemed to rest on the succession of a son to his father; much, on the contrary, on the adoption of a stranger of tried capacity guided by the choice of the actual ruler. But in the lapse of centuries the imperial power had become absolute. Especially in the successors of Constantine, and in the city to which he had given his name and chosen for the home of his empire, not a shadow of the old Roman freedom remained. One after another the successful general or the adventurer in some court intrigue supplanted or murdered a predecessor, and ascended the throne, but with undiminished prerogatives. Great was the contrast in all the new kingdoms at whose birth the influence of the Church presided. There the kings all sat by family descent, in which, however, was involved a free acceptance on the part of their people. The bishops who had had so large a part in the foundation of the several kingdoms had a recognised part in their future government. Holding one faith, and educated in the law of the Romans, and joined on to the preceding ages by their mental culture as well as their belief, they contributed to these kingdoms a stability and cohesion which were wanting to the Teuton invaders in themselves. They incessantly preached peace as a religious necessity to those tribes which had been as ready to consume each other as to divide the spoils of their Roman subjects. This united phalanx of bishops in Gaul conquered in the end even the excessive degeneracy, self-indulgence, and cruelty of the Merovingian race. Thanks to their perpetual efforts, while the policy of a Clovis made a France, the wickedness of his descendants did not destroy it, but only themselves, and caused a new family to be chosen wherein the same tempered government might be carried on.

It is remarkable that while the Byzantine emperors, from the extinction of the western empire, were using their absolute power to meddle with the doctrine of the Church which Constantine acknowledged to be divine, and to fetter its liberty which he acknowledged to be unquestionable, the Popes from that very time were through the bishops, to whom they were the sole centre in so many changes and upheavals, constructing the new order of things. Through them the Church maintained her own liberty, and allied with it a civil liberty which the East had more and more surrendered.

In the East, the Church in time was younger than the empire; in the West, she preceded in time these newly formed monarchies. Amid the universal overthrow which the invaders had wrought she alone stood unmoved. The heresy which had so threatened her disappeared. On Goths, and Franks, and Saxons, and Alemans, she was free to exercise her divine power.[1] It is in that sixth century of tremendous revolutions

---

[1] See Kurth, ii. 25-6.

that she laid the foundation of the future European society. Byzantium was descending to Mahomet while Rome was forecasting the Christian commonwealth of Charles the Great. In the Rome of Constantine, while the old civilisation had accepted her name, the old pagan principles had continually impeded her action. The civil rulers especially had harked back after the power of the heathen Pontifex Maximus; but in these new peoples who were not yet peoples, but only the unformed matter (*materia prima*) out of which peoples might be made, the Church was free to put her own ideal as a *form* within them. They had the rudiments of institutions, which they trusted her to organise. They placed her bishops in their courts of justice, in their halls of legislation. The greatest of their conquerors in the hour of his supreme exaltation, which also was received from the Pope, was proud to be vested by her in the dalmatic of a deacon.

Of this new world St. Gregory, in his desolated Rome, stood at the head.

There is yet another aspect of this wonderful man which we have to consider. We possess about 850 of his letters. If we did but possess the letters of his sixty predecessors in the same relative proportion as his, the history of the Church for the five centuries preceding him, instead of being often a blank, would present to us the full lineaments of truth. The range of his letters is so great, their detail so minute, that they illuminate his time and enable us to form a mental picture, and follow faithfully that pontificate of fourteen years, incessantly interrupted by cares and anxieties for the preservation of his city, yet watching the beginnings and strengthening the polity of the western nations, and counterworking the advances of the eastern despotism. The divine order of greatness is, we know, to do and to teach. Few, indeed, have carried it out on so great a scale as St. Gregory. The mass of his writing preserved to us exceeds the mass preserved to us from all his predecessors together, even including St. Leo, who with him shares the name of Great, and whose sphere of action the mind compares with his. If he became to all succeeding times an image of the great sacerdotal life in his own person, so all ages studied in his words the pastoral care, joining him with St. Gregory of Nazianzum and St. Chrysostom. The man who closed his life at sixty-four, worn out not with age, but with labour and bodily pains, stands, beside the learning of St. Jerome, the perfect episcopal life and statesmanship of St. Ambrose, the overpowering genius of St. Augustine, as the fourth doctor of the western Church, while he surpasses them all in that his doctorship was seated on St. Peter's throne. If he closes the line of Fathers, he begins the period when the Church, failing to preserve a rotten empire in political existence, creates new nations; nay, his own hand has laid for them their foundation-stones, and their nascent polity bears his manual inscription, as the great campanile of St. Mark wears on its brow the words, *Et Verbum caro factum est.* These were the words which St. Gregory wrote as the bond of their internal cohesion, as the source of their greatness, permanence, and liberty upon the future monarchies of Europe.

What mortal could venture to decide which of the two great victories allowed by Gibbon to the Church is the greater? But we at least are the children of the second. It was wrought in secrecy and unconsciousness, as the greatest works of nature and of grace are wrought, but we know just so much as this, that St. Gregory was one of its greatest artificers. The Anglo-Saxon race in particular, for more than a thousand years, has celebrated the Mass of St. Gregory as that of the Apostle of England. Down to the disruption of the sixteenth century, the double line of its bishops in Canterbury and York, with their suffragans, regarded him as their founder, as much as the royal line deemed itself to descend from William the Conqueror. If Canterbury was Primate of all England and York Primate of England, it was by the appointment of Gregory. And the very civil constitution of England, like the original constitutions of the western kingdoms in general, is the work in no small part of that Church which St. Augustine carried to Ethelbert, and whose similar work in Spain Gibbon has acknowledged. Under the Norman oppression it was to the laws of St. Edward that the people looked back. The laws of St. Edward were made by the bishops of St. Gregory.

How deeply St. Gregory was impressed with the conviction of his own vocation to be the head of the whole Church we have seen in his own repeatedly quoted words.[1] What can a Pope claim more than the attribution to himself as Pope of the three great words of Christ spoken to Peter? Accordingly, all his conduct was directed to maintain every particular church in its due subordination to the Roman Church, to reconcile schismatics to it, to overcome the error and the obstinacy of heretics. Again, since all nations have been called to salvation in Christ, St. Gregory pursued the conversion of the heathen with the utmost zeal. When only monk and cardinal deacon, he had obtained the permission of Pope Pelagius to set out in person as missionary to paganised Britain. He was brought back to Rome after three days by the affection of the people, who would not allow him to leave them. When the death of Pope Pelagius placed him on the papal throne, he did not forget the country the sight of whose enslaved children had made them his people of predilection.

With regard to the churches belonging to his own patriarchate, a bishop in each province, usually the metropolitan, represented as delegate the Roman See. To these, as the symbol of their delegated authority as his *vicarii*, Gregory sent the pallium. All the bishops of the province yielded them obedience, acknowledged their summons to provincial councils. A hundred years before Pope Symmachus had begun the practice of sending the pallium to them, but Gregory declined to take the gifts which it had become usual to take on

---

[1] See in the *Kirchen-lexicon* of Card. Hergenröther the article on Gregory I., vol. v., p. 1079.

receiving it. St. Leo, fifty years before Symmachus, had empowered a bishop to represent him at the court of the eastern emperor, and had drawn out the office and functions of the nuncio. Like his great predecessor, St. Gregory carefully watched over the rights of the Primacy. Upon the death of a metropolitan, he entrusted during the vacancy the visitation of the churches to another bishop, and enjoined the clergy and people of the vacant see to make a new choice under the superintendence of the Roman official. The election being made, he carefully examined the acts, and, if it was needed, reversed them. As he required from the metropolitans strict obedience to his commands, so he maintained on the one hand the dependence of the bishops on their metropolitans, while on the other he protected them against all irregular decisions of the metropolitan. He carefully examined the complaints which bishops made against their metropolitan; and when bishops disagreed with each other, and their disagreement could not be adjusted by the metropolitan, he drew the decision to himself.

Gregory also held many councils in Rome which passed decisions upon doctrine and discipline. We may take as a specimen that which he held in the Lateran Church on the 5th April, 601,[1] with twenty-four bishops and many priests and deacons. It is headed: "Gregory, bishop, servant of the servants of God, to all bishops". The Pope says that his own

---

[1] See Hefele, *Concilien-geschichte*, iii., p. 56; St. Gregory, ii., p. 1294; Mansi, x., p. 486.

government of a monastery had shown him how necessary it was to provide for their perpetual security: "Since we have come to the knowledge that in very many monasteries the monks have suffered much to their prejudice and grievance from bishops . . . we therefore, in the name of our Lord Jesus Christ, and by the authority of the blessed Peter, prince of the Apostles, in whose place we preside over this Church, forbid that henceforth any bishop or layman, in respect of the revenues, goods, or charters of monasteries, the cells or buildings belonging to them, do in any manner or upon any occasion diminish them, or use deceit or interference". If there be a contest whether any property belong to the church of a bishop or to a monastery, arbitrators shall decide. If an abbot dies, no stranger, but one of the same community, must be chosen by the brethren, freely and concordantly, for his successor. If no fitting person is found in the monastery itself, the monks are to provide that one be chosen from another monastery. In the abbot's lifetime no other superior may be set over the monastery, except the abbot have committed transgressions punishable by the canons. Against the will of the abbot no monk may be chosen to be set over another monastery or receive holy orders. The bishop may not make an inventory of the goods of the monastery, nor mix himself, even after the abbot's death, in the concerns of the monastery; he may hold no public mass in the monastery, that there be no meeting of people, or women, there; he may set up no pulpit there, and without the consent of the abbot make no regulation, and employ no monk for any church service.

All the bishops answered: "We rejoice in the liberties of the monks, and confirm what your Holiness has set forth as to this".

As metropolitan of the particular Roman province, Gregory was equally active. The political circumstances of Italy had exerted the most prejudicial effect on the Church. Ecclesiastical life was impaired. The discipline both of monks and clergy was weakened. Bishops had become negligent in their duties; many churches orphaned or destroyed. But at the end of his pontificate things had so improved that he might well be termed the reformer of Church discipline. He watched with great care over the conduct and administration of the bishops. In this the officers called *defensors*, that is, who administered the patrimony of of the Church in the different provinces, helped him greatly in carrying out his commands. In the war with the Lombards, many episcopal sees had been wasted, and many of their bishops expelled. Gregory provided for them, either in naming them visitors of his own, or in calling in other bishops to their support. He rebuilt many churches which had been destroyed. He carefully maintained the property of churches: he would not allow it to be alienated, except to ransom captives or convert heathens. The Roman Church had then large estates in Africa, Gaul, Sicily, Corsica, Dalmatia, and especially in the various provinces of Italy. These were called the Patrimony of Peter. They consisted in lands, villages, and flocks. In the management of these Gregory's care did not disdain the minutest supervision. His strong sense of justice did not prevent his being a merciful landlord, and especially he cared for the peasantry and cultivators of the soil.

The monastic life which in his own person he had so zealously practised, as Pope he so carefully watched over that he has been called the father of the monks. He encouraged the establishment of monasteries. Many he built and provided for himself out of the Roman Church's property. Many which wanted for maintenance he succoured. He issued a quantity of orders supporting the religious and moral life of monks and nuns. He invited bishops to keep guard over the discipline of monasteries, and blamed them when transgressions of it came to light. But he also protected monasteries from hard treatment of bishops, and, according to the custom of earlier Popes, exempted some of them from episcopal authority.

In restoring schismatics to unity he was in general successful. He wrought such a union among the bishops of Africa that Donatism lost influence more and more, and finally disappeared. He dealt with the obstinate Milanese schism which had arisen out of the treatment of the Three Chapters. He won back a great part of the Istrians. He had more trouble with the two archbishops of Constantinople, John the Faster and Cyriacus; and his former friend the emperor Mauritius turned against him, so that he welcomed the accession of Phocas, as a deliverance of the Church from unjust domination. The unquestioning loyalty with which, as a civil subject, he welcomed this accession has been unfairly used against him. As first of all the civil dignitaries of the empire he could only accept what had been done at Constantinople. But in all his fourteen years neither the difficulty of circumstances nor the consideration of persons withheld him from carrying out his resolutions with a patience and a firmness only equalled by gentleness of manner. From beginning to end he considered himself, and acted, as set by God to watch over the maintenance of the canons, the discipline enacted by them, and so doing to perfect by his wisdom as well as to temper by his moderation the vast fabric of the Primacy as it had grown itself, and nurtured in its growth the original constitution of the Church during nearly six hundred years.

We may now say a few words upon the Primacy itself as exerted by St. Leo at the Council of Chalcedon, and the Primacy as exerted by St. Gregory in the fourteen years from 590 to 604; also on the interval between them, and the relative position of the bishop of Constantinople to Leo in the person of Anatolius, and to Gregory in the person of John the Faster. We see at once that the intention which Leo discerned in Anatolius, which he sternly reprehended and summarily overthrew, has been fully carried out by John the Faster, who, in documents sent to the Pope himself for revision, as superior, terms himself ecumenical patriarch. Who had made him first a patriarch and then ecumenical? The emperor alone. He is so called in the laws of Justinian. The 140 years from Leo to Gregory are filled with the continued rise of the Bishop of Nova Roma under the absolute power of the emperor. He has succeeded not only in taking precedence of the legitimate patriarchs of Alexandria and Antioch; he has more than once stripped of their rights the metropolitans and bishops subject to the great see of the East, and himself consecrated at Constantinople a patriarch of Antioch by order of the emperor of the day. This Acacius did, humbly begging the Pope's pardon for such a transgression of the due order and hierarchy, and repeating the offence against the Nicene order and constitution on the first opportunity. In the same way he has interfered with the elections at Alexandria. We learn from the instruction given by Pope Hormisdas to his legates that all the eastern bishops when they came to Constantinople obtained an audience of the emperor only through the bishop of Constantinople. The Pope carefully warns his legates against submitting to this pretension. Pope Gelasius told the bishop in his day that his see had no ecclesiastical rank above that of a simple bishop. We laugh, he said, at the pretension to erect an apostolical throne upon an imperial residence. But, in the meantime, Constantinople has become the head of all civil power. The emperor of the West has ceased to be. The Roman senate, at the bidding of a Herule commander of mercenaries, has sent back even the symbols of imperial rank to the eastern emperor; and in return Zeno has graciously made Odoacer patricius of Rome, with the power of king, until Theodorick was ready to be rewarded with the possession of Italy for services rendered to the eastern monarch, with the purpose likewise of diverting his attention from Nova Roma. Therefore, in spite of the submission rendered by all the East, the bishops, the court, the emperor, and by Justinian himself; in spite, also, of two bishops successively degraded by an emperor, the bishop of Constantinople ever advances. The law of Justinian, which acknowledges the Pope as first of all bishops in the world, and gives him legal rank as such, makes the bishop of the new capital the second. Presently Justinian becomes by conquest immediate sovereign of Rome. The ancient queen and maker of the empire is humbled in the dust by five captures; is even reduced to a desert for a time; and when a portion of her fugitive citizens comes back to the abandoned city, a Byzantine prefect rules it with absolute power. A Greek garrison, the badge of Rome's degradation, supports his delegated rule. Presently the seat of that rule is for security transferred to Ravenna, and Rome is left, not merely discrowned, but defenceless. All the while the bishop of Constantinople is seated in the pomp of power at the emperor's court; within the walls of the eastern capital his household rivals that of the emperor; in certain respects the public worship gives him a homage greater than that accorded to the absolute lord of the East. He reflects with satisfaction that the one person in the West who can call his ministration to account is exposed to the daily attacks of barbarians: is surrounded with palaces whose masters are ruined, and which are daily dropping into decay. The Pope, behind the crumbling walls of Aurelian, shudders at the cruelties practised on his people: the bishop of Constantinople, by terming himself œcumenical, announces ostentatiously that he claims to rule all his brethren in the East—that he is supreme judge over his brother patriarchs. One only thing he does not do: he claims no power over the Pope himself; he does not attempt to revise his administration in the West. He acknowledges his primacy, seated as it is in a provincial city, pauperised, and decimated with hunger and desertion.

In this interval the Pope has seen seven emperors pass like shadows on the western throne, and their place taken first by an Arian Herule and then by an Arian Goth. Herule and Goth disappear, the last at the cost of a war which desolates Italy during twenty years, and casts out, indeed, the Gothic invader and confiscator of Italy, but only to supply his place by the grinding exactions of an absent master, followed immediately by the inroad of fresh savages, far worse than the Goth, under whose devastation Italy is utterly ruined. Whatever portion of dignity the old capital of the world lent to Leo is utterly lost to Gregory. It has been one tale of unceasing misery, of terrible downfal to Rome, from Genseric to Agilulf. It may seem to have been suspended during the thirty-three years of Theodorick, but it was the iron force of hostile domination wielded by the gloved hand. When the Goth was summoned to depart, he destroyed ruthlessly. The rage of Vitiges casts back a light upon the mildness of Theodorick; the slaughters ordered by Teia are a witness to Gothic humanity. No words but those of Gregory himself, in applying the Hebrew prophet, can do justice to the temporal misery of Rome. The Pope felt himself silenced by sorrow in the Church of St. Peter, but he ruled without contradiction the Church in East and West. Not a voice is heard at the time, or has come down to posterity, which accuses Gregory of passing the limits of power conceded to him by all, or of exercising it otherwise than with the extremest moderation.

Disaster in the temporal order, continued through five generations, from Leo to Gregory, has clearly brought to light the purely spiritual foundation of the papal power. If the attribution to the Pope of the three great words spoken by our Lord to St. Peter, made to Pope Hormisdas by the eastern bishops and emperor, does not prove that they belong to the Pope and were inherited by him from St. Peter, what proof remains to be offered? If the attribution is so proved, what is there in the papal power which is not divinely conferred and guaranteed? Neither the first Leo, nor the first Gregory, nor the seventh Gregory, nor the thirteenth Leo, ask for more; nor can they take less.

If St. Gregory exercised this authority in a ruined city, over barbarous populations which had taken possession of the western provinces, over eastern bishops who crouched at the feet of an absolute monarch, over a rival who, with all the imperial power to back him, did not attempt to deny it, how could a greater proof of its divine origin be given?

In this respect boundless disaster offers a proof which the greatest prosperity would have failed to give. Not even a Greek could be found who could attribute St. Gregory's authority in Rome to his being bishop of the royal city. The barbarian inundation had swept away the invention of Anatolius.

But this very time was also that in which the heresy whose leading doctrine was denial of the Godhead of the Church's founder came from a threatening of supremacy to an end. In Theodorick Arianism seemed to be enthroned for predominance in all the West. His civil virtues and powerful government, his family league of all the western rulers,—for he himself had married Adolfleda, sister of Clovis, and had given one daughter for wife to the king of the Vandals in Africa, and another to the king of the Visigoths in France,—was a gage of security. In Gregory's time the great enemy has laid down his arms. He is dispossessed from the Teuton race in its Gallic, Spanish, Burgundian, African settlements. Gregory, at the head of the western bishops who in every country have risked life for the faith of Rome, has gained the final victory. One only Arian tribe survives for a time, ever struggling to possess Rome, advancing to its gates, ruining its Campagna, torturing its captured inhabitants, but never gaining possession of those battered walls, which Totila in part threw down and Belisarius in piecemeal restored. And Gregory, too, is chosen to stop the Anglo-Saxon revel of cruelty and destruction, which has turned Britain from a civilised land into a wilderness, and from a province of the Catholic Church to paganism, from the very time of St. Leo. Two tribes were the most savage of the Teuton family, the Saxon and the Frank. The Frank became Catholic, and Gregory besought the rulers of the converted nation to help his missionaries in their perilous adventure to convert the ultramarine neighbours, still savage and pagan. He also ordered their chief bishop to consecrate the chief missionary to be archbishop of the Angles. As there was a Burgundian Clotilda by the side of Clovis, there was a Frankish Bertha by the side of Ethelbert; and these two women have a glorious place in that second great victory of the Church. The Visigoth and Ostrogoth with their great natural gifts could not found a kingdom. Their heresy deprived the Father of the Son, and they were themselves sterile. Those who denied a Divine Redeemer were not likely to convert a world.

But all through Gregory's life the Byzantine spirit of encroachment was one of his chief enemies. The claim of its bishop to be ecumenical patriarch stopped short of the Primacy. But one after another the bishops of that see sought by imperial laws to detach the bishops of Eastern Illyria from their subjection to the western patriarchate. Their nearness to Constantinople, their being subjects of the eastern emperor, helped this encroachment.

It would appear also that in Gregory's time—a

hundred years after Pope Gelasius had put the bishop of the imperial city in remembrance that he had been a suffragan to Heraclea—the legislation of Justinian had succeeded in inducing the Roman See to acknowledge that bishop as a patriarch. His actual power had gone far beyond. There can be no doubt that, while the Pope had become legally the subject of the eastern emperor, the bishop of Constantinople had become in fact the emperor's ecclesiastical minister in subjugating the eastern episcopate. The Nicene episcopal hierarchy subsisted indeed in name. To the Alexandrian and Antiochene patriarchs two had been added—one at Jerusalem, the other at Constantinople. But the last was so predominant—as the interpreter of the emperor's will—that he stood at the head of the bishops in all the realm ruled from Constantinople over against the Pope as the head of the western bishops in many various lands.

The bishops were in Justinian's legislation everywhere great imperial officers, holding a large civil jurisdiction, especially charged with an inspection of the manner in which civil governors performed their own proper functions; most of all, the patriarchs and the Pope.

But that episcopal autonomy—if we may so call it—under the presidence of the three Petrine patriarchs, which was in full life and vigour at the Nicene Council, which St. Gregory still recognised in his letter to Eulogius, was greatly impaired. While barbaric inundation had swept over the West, the struggles of the Nestorian and Eutychean heresies, especially in the two great cities of Alexandria and Antioch, had disturbed the hierarchy and divided the people which the master at Constantinople could hardly control. That state of the East which St. Basil deplored in burning words—which almost defied every effort of the great Theodosius to restore it to order—had gone on for more than two hundred years. The Greek subtlety was not pervaded by the charity of Christ, and they carried on their disputes over that adorable mystery of His Person in which the secret of redeeming power is seated, with a spirit of party and savage persecution which portended the rise of one who would deny that mystery altogether, and reduce to a terrible servitude those who had so abused their liberty as Christians and offered such a scandal to the religion of unity which they professed.

From St. Sylvester to St. Leo, and, again, from St. Leo to St. Gregory, the effort of the Popes was to maintain in its original force the Nicene constitution of the Church. Well might they struggle for the maintenance of that which was a derivation from their own fountain-head—"the administration of Peter"[1]—during the three centuries of heathen persecution by the empire. It was not they who tightened the exercise of their supreme authority. The altered condition of the times, the tyranny of Constantius and Valens, the dislocation of the eastern hierarchy, the rise of a new bishop in a new capital made use of by an absolute sovereign to control that hierarchy, a resident council at Constantinople which became an "instrument of servitude" in the emperor's hands to degrade any bishop

[1] S. Siricius, *Ep.*

at his pleasure and his own patriarch when he was not sufficiently pliant to the master,—these were among the causes which tended to bring out a further exercise of the power which Christ had deposited in the hands of His Vicar to be used according to the needs of the Church. No one has expressed with greater moderation than St. Gregory the proper power of his see, in the words I have quoted above : [1] "I know not what bishop is not subject to the Apostolical See, if any fault be found in bishops. But when no fault requires it, all are equal according to the estimation of humility." In Rome there is no growth by aid of the civil power from a suffragan bishop to an universal Papacy. The Papacy shows itself already in St. Clement, a disciple of St. Peter's, "whose name is written in the book of life," [2] and who, invoking the Blessed Trinity, affirms that the orders emanating from his see are the words of God Himself.[3] This is the ground of St. Gregory's moderation ; and whatever extension may hereafter be found in the exercise of the same power by his successors is drawn forth by the condition of the times, a condition often opposed to the inmost wishes of the Pope. Those

---

[1] P. 308.

[2] Philippians iv. 3.

[3] See St. Clement's epistle, sec. 59. "Receive our counsel and you shall not repent of it. For, as God liveth, and as the Lord Jesus Christ liveth, and the Holy Spirit, and the faith and the hope of the elect, he who performs in humility, with assiduous goodness, and without swerving, *the commands and injunctions of God*, he shall be enrolled and esteemed in the number of those saved through Jesus Christ, through whom be glory to Him for ever and ever. Amen. But if any disobey *what has been ordered by Him through us*, let them know that they will involve themselves in a fall, and no slight danger, but we shall be innocent of this sin."

are evil times which require "a thousand bishops rolled into one" to oppose the civil tyranny of a Hohenstaufen, the violence of barbarism in a Rufus, or the corruption of wealth in a Plantagenet.

Between St. Peter and St. Gregory, in 523 years, there succeeded full sixty Popes. If we take any period of like duration in the history of the world's kingdoms, we shall find in their rulers a remarkable contrast of varying policy and temper. Few governments, indeed, last so long. But in the few which have so lasted we find one sovereign bent on war, another on peace, another on accumulating treasure, another on spending it; one given up to selfish pleasures, here and there a ruler who reigns only for the good of others. But in Gregory's more than sixty predecessors there is but one idea: "Thou art Peter, and upon this rock I will build My Church, and the gates of hell shall not prevail against it," is the compendious expression of their lives and rule. For this St. Clement, who had heard the words of his master, suffered exile and martyrdom in the Crimea. For this five Popes, in the decade between 250 and 260, laid down their lives. The letter of St. Julius to the Eusebian prelates is full of it. St. Leo saw the empire of Rome falling around him, but he is so possessed with that idea that he does not allude to the ruin of temporal kingdoms. St. Gregory trembles for the lives of his beleaguered people, but he does not know the see which is not subject to the Apostolic See. In weakness and in power, in ages of an ever varying but always persistent adversity, in times of imperial patronage, and, again, under heretical domination, the mind of every Pope is full of this idea. The strength or the weakness of individual character leaves it untouched. In one, and only one, of all these figures his dignity is veiled in sadness. Pope Vigilius at Constantinople, in the grasp of a despot, and with the stain of an irregular election never effaced from his brow, is still conscious of it, still has courage to say, "You may bind me, but you will not bind the Apostle St. Peter". Six hundred years after St. Gregory, when accordingly the succession of Popes had been rather more than doubled, I find the biographer of Innocent III. thus commenting on his election in 1198: "The Church in these times ever had an essential preponderance over worldly kingdoms. Resting on a spiritual foundation, she had in herself the vigour of immaterial power, and maintained in her application of it the superiority over merely material forces. She alone was animated by a clearly recognised idea, which never at any time died out of her. For its maintenance and actuation were not limited to the person of a Pope, who could only be the representative, the bearer, the enactor, for the world of this idea in its fullest meaning. If here and there a particular personality seemed unequal to the carrying out such a charge, the force of the idea did not suffer any defect through him. Most papal governments were very short in their duration. This itself was a challenge to those whose life was absorbed in that of the Church to place at its head a man whose ability, enlightened and guided by strength of will, afforded a secure assurance for the exercise of an universal charge. From the clear self-consciousness of the Church in this respect proceeded that firm pursuance of a great purpose distinctly perceived. It met with no persistent or wisely conducted resistance on the part of the temporal power. On one side all rays had their focus in one point. In temporal princes the rays were parted. Few of these showed in their lives a purpose to which all their acts were made consistently subordinate. As circumstances swayed them, as the desire of the moment led them away, they threw themselves, according to their personal inclinations, with impetuous storm and violence upon the attainment of their wishes. They had to yield in the end to the power of the Church, slower, indeed, but continuous, pursued with superiority of spirit, moreover with the firm conviction of guidance from above, and of the special protection from this inseparable, and so attaining its mark. One only royal race ventured on a contest with the Church for supremacy; for one only, the Hohenstaufen, were conscious of a fixed purpose. They encountered a direct struggle with the Church; but the conflict issued to the honour of the Church. The Popes who led it came out of it with a renown in the world's history, which without that conflict they would never have so gloriously attained. If we look from these events before and afterwards upon the ages, and see how the institution of the Papacy outlasts all other institutions in Europe, how it has seen all States come and go, how in the endless change of human things it alone remains unchanged, ever with the same spirit, can we then wonder if many look up to it as the Rock unmoved amid the roaring billows of centuries?" And he adds in a note, " This is not a polemical statement, but the verdict of history".[1]

The time of St. Gregory in history bore the witness of six centuries; the time of Innocent III. of twelve; the time of Leo XIII. bears that of more than eighteen centuries to the consideration of this contrast between the natural fickleness of men and of lives of men, shown from age to age, and the persistence, on the other hand, of one idea in one line of men. The eighteen centuries already past are yet only a part of an unknown future. But to construct such a Rock amid the sea and the waves roaring in the history of the nations reveals an abiding divine power. It leaves the self-will of man untouched, yet sets up a rampart against it. The explanation attempted three hundred and fifty years ago of an imposture or an usurpation is incompatible with the clearness of an idea which is carried out persistently through so many generations. Usurpations fall rapidly. But in this one case the divine words themselves contain the idea more clearly expressed than any exposition can express it. The King delineates His kingdom as none but God can; it must also be added that He maintains it as none but God can maintain.

We may return to St. Gregory's own time, and note the unbroken continuity of the Primacy from St. Peter himself. It is a period of nearly six hundred years from the day of Pentecost. Just in the middle comes

---

[1] Hurter's *Geschichte Papst Innocenz des Dritten*, i. 85-7.

the conversion of Constantine. Before it Rome is mainly a heathen city, the government of which bears above all things an everlasting enmity against any violation of the supreme pontificate annexed by the provident Augustus to the imperial power, and jealously maintained by every succeeding emperor. To suffer an infringement of that pontificate would be to lose the grasp over the hundred varieties of worship allowed by the State. Yet when Constantine acknowledged the Christian faith, the names of St. Peter and St. Paul were in full possession of the city, so far as it was Christian. They were its patron-saints. Every Christian memory rested on the tradition of St. Peter's pontifical acts, his chair, his baptismal font, his dwelling-place, his martyrdom. The impossibility of such a series of facts taking possession of a heathen city during the period antecedent to Constantine's victory over Maxentius, save as arising from St. Peter's personal action at Rome, is apparent.

In the second half of this period, from Constantine to St. Gregory, the civil pre-eminence of Rome is perpetually declining. The consecration of New Rome as the capital of the empire, in 330, by itself alone strikes at it a fatal blow. Presently the very man who had reunited the empire divided it among his sons, and after their death the division became permanent. Valentinian I., in 364, whether he would or not, was obliged to make two empires. From the death of Theodosius, in 395, the condition of the western empire is one long agony. The power of Constantinople continually increases. At the death of Honorius, in 423, the eastern emperor becomes the over-lord of the western. During fifty years Rome lived only by the arm of two semi-barbarian generals, Stilicho and Aetius. Both were assassinated for the service; and in the boy Romulus Augustulus a western emperor ceased to be, and the senate declared that one emperor alone was needed. After fifty years of Arian occupation, the Gothic war ruined the city of Rome. In Gregory's time it had ceased to be even the capital of a province. Its lord dwelt at Constantinople; Rome was subject to his exarch at Ravenna.

Yet from Constantine and the Nicene Council the advance of Rome's Primacy is perpetual. In Leo I. it is universally acknowledged. At the fall of the western empire Acacius attempts his schism. He is supported while living by the emperor Zeno, and his memory after his death by the succeeding emperor Anastasius, who reigned for twenty-seven years, longer than any emperor since Augustus had reigned over the whole empire. All the acts of these two princes show that they would have liked to attach the Primacy to their bishop at Constantinople. Anastasius twice enjoyed the luxury of deposing him through the resident council. But Anastasius died, and the result of the Acacian schism was a stronger confession of the Roman Primacy made to Pope Hormisdas, the subject of the Arian Theodorick, by the whole Greek episcopate, than had ever been given before. The sixth century and the reign of Justinian completed the destruction of the civil state of Rome; and the Primacy of its bishop, St. Gregory, was more than ever acknowledged.

Not a shadow of usurpation or of claim to undue power rested upon that unquestioned Primacy which St. Gregory exercised. While he thought the end of the world was at hand, while he watched Rome perishing street by street, he planted unconsciously a western Christendom in what he supposed all the time to be a perishing world. Civil Rome was not even a provincial capital; spiritual Rome was the acknowledged head of the world-wide Church.

I know not where to find so remarkable a contrast and connection of events as here. Temporal losses, secular ambitions, episcopal usurpations, violent party spirit, schism and heresy in the great eastern patriarchates, and amid it all the descent of the Teutons on the fairest lands of the western empire, the establishment of new sovereignties in Spain, Gaul, and Italy, under barbarians who at the time of their descent were Arian heretics, and afterwards became Catholic, with the result that Gregory has to keep watch within the walls of Rome for a whole generation against the Lombard, still in unmitigated savagery and unabated heresy, and that the world-wide Church acknowledges him for her ruler without a dissenting voice. The "Servant of the servants of God" chides and corrects the would-be "ecumenical patriarch," who has risen since Constantine from the suffragan of a Thracian city to be bishop of Nova Roma and right hand of the emperor; who has deposed Alexandria from the second place and Antioch from the third, but cannot take the first place from the See of Peter. The perpetual ambition of the bishops of Nova Roma, the perpetual fostering of that ambition for his own purpose by the emperor, only illustrates more vividly the inaccessible dignity which both would fain have transferred to the city of Constantine, but were obliged to leave with the city of Peter. As the forum of Trajan sinks down stone by stone, the kings of the West are preparing to flock in pilgrimage to the shrine of Peter. This was the answer which the captives in the forum made to the deliverer of their race.

There is nothing like this elsewhere in history. Constantine, Valens, Theodosius, Justinian, and, no less, Alaric and Ataulph, Attila and Genseric, Theodorick and Clovis, Arius, Nestorius, Eutyches, as well as St. Athanasius, St. Basil, St. Ambrose, St. Chrysostom, St. Augustine, St. Cyril, and, again, Dioscorus, Acacius, and a multitude of the most opposing minds and beliefs which these represent, contribute, in their time and degree, for the most part unconsciously, and many against their settled purpose, to acknowledge this Primacy as the Rock of the Church, the source of spiritual jurisdiction, the centre of a divine unity in a warring world. In St. Gregory we see the power which has had antecedents so strange and concomitants so repulsive deposited in the hands of a feeble old man who is constantly mourning over the cares in which that universal government involves him, while the world for evermore shall regard him as the type and standard of the true spiritual ruler, who calls himself, not Ecumenical Bishop, but Servant of the servants of God. It is a title which his successors will take from his hand and keep for ever as the badge of the Primacy which it illustrates, while it serves as the seal of its acts of power. He calls himself servant just when he is supreme.

In St. Gregory the Great, the whole ancient world, the Church's first discipline and original government, run to their ultimate issue. In him the patriarchal system, as it met the shock of absolute power in the civil sovereign, and the subversion of the western empire by barbarous incursions, accompanied by the establishment of new sovereignties and the foundation of a new Rome, the rival and then the tyrant of the old Rome, receives its consummation. The medieval world has not yet begun. The spurious Mahometan theocracy is waiting to arise. In the midst of a world in confusion, of a dethroned city falling into ruins, the successor of St. Peter sits on an undisputed spiritual throne upon which a new world will be based in the West, against which the Khalifs of a false religion will exert all their rage in the East and South, and strengthen the rule which they parody. A new power, which utterly denies the Christian faith, which destroys hundreds of its episcopal sees and severs whole countries from its sway, will dash with all its violence against the Rock of Peter, and finally will have the effect of making the bishop who is there enthroned more than ever the symbol, the seat, and the champion of the Kingdom of the Cross.

INDEX.

INDEX.

Acacius, bishop of Constantinople, 471-489, 65; his conduct to the year 482, 66; induces Zeno to publish a formulary of doctrine, 70; deposed by Pope Felix, 75; rejects the Pope's sentence, 83; attempts superiority over the eastern patriarchates, 84-86; position taken up by him against the Pope, 84-91; dies after five years of excommunication in 489, defying the Pope, 83; his name erased from the diptychs, 168; summary of his conduct and aims, 174-6

Agapetus, Pope, his accession, 202; confirms all his old rights to the Primate of Carthage, 203; confirms Justinian's profession of faith, at the emperor's request, 204; goes to Constantinople, deposes Anthimus and consecrates Mennas patriarch, 205

Agnostics, generated by schismatics, 5

Alexandria and Antioch, fearful state of their patriarchates, 184; the vast difference between their patriarchs and the Primacy, 185

Anastasius II., Pope, 496-8, 120; his letter to the emperor asserts that as the imperial secular dignity is preeminent in the whole world, so the Principate of St. Peter's See in the whole Church, 120; both are divine delegations, 121; writes to Clovis upon his conversion, 122; anticipates the great results to follow from it, 123

Anastasius, eastern emperor in 491, made emperor when a *Silentiarius* in the court, 518, 83; summary of his reign in the "libellus synodicus," 100-1; four Popes—Gelasius, Anastasius, Symmachus, and Hormisdas—have to deal with him, 102; tries to prevent the election of Pope Symmachus, 129; he is obliged to allow the Roman See not to be judged, 143; he deposes Euphemius, and puts Macedonius in his stead at Constantinople, 143; exalts Timotheus to the see of Constantinople, 148; fills the eastern patriarchal sees with heretics, 149; being pressed by Vitalian, betakes himself to Pope Hormisdas, 150; receives his conditions, except those concerning Acacius, 159; his treachery and cruelty, 160; his sudden death, 162

Anatolius, bishop of Constantinople, crowns the emperor Leo I., dies in 458, 64; his ambition seen and checked by St. Leo, 60; is to Leo what John the Faster is to Gregory, 307

Anicius Olybrius, Roman emperor, 20

Anthemius, Roman emperor, 18

Arianism, propagated among the Goths by the emperor Valens, 49; communicated by them to the Teuton tribes, 29; prevalent throughout the West, 50; fails in the Vandal, Visigothic, Burgundian, and Ostrogothic kingdoms, 327-9

Aspar, Arian Goth, makes Leo I. emperor, and is slain by him, 62

Ataulph, marries Galla Placidia, his judgment upon the Goths and Romans, 43

Avitus, St., bishop of Vienne, in Gaul, his character of Acacius, 93; his letter to Clovis on his conversion, 124; urges his duty to propagate the faith in the peoples around him, 126; writes to the Roman senate that the cause of the Bishop of Rome is not one bishop but that of the Episcopate itself, 140

Avitus, Roman emperor, 13

Augustine, St., the great victory of the Church which he did not foresee, 57

Baronius, quoted, 76, 79, **202**, **207**

Basiliscus, **usurper, first of the theologising emperors**, 46

Belisarius, reconquers Northern Africa, 199; begins the Gothic war, and enters Rome, 205; deposes Pope Silverius, 207; defends Rome against Vitiges, 210; captures Rome the third time, 207

Benedict, St., his monastery at Monte Cassino destroyed by the Lombards, 290; his Order has its chief seat for 140 years at St. John Lateran, **290**; rebukes and subdues Totila, 215

Byzantium, **the over-lordship of its emperor acknowledged, 18, 23; the succession to its throne, 61; its constitution under Justinian contrasted with the medieval constitution of England, 250**

Cassiodorus, his letter as Prætorian prefect to Pope John II., **195**

Church, Catholic, **its two great victories, 5, 25; attested and described by** Gibbon, 325

Civiltà Cattolica, quoted, 103, 104, 128

Constantinople, its seven bishops who follow Anatolius, 180; submission of its bishop, clergy, emperor, and nobles to Pope Hormisdas, 187; service of its cathedral under Justinian, 244; growth of its bishop from St. Leo to St. Gregory, 342; all **the** work of the imperial power, 344; perpetual encroachment of its bishops, 348, 359

Cyprian, St., quoted, "**De Unitate Ecclesiæ**," 3

Dante, quoted, 184; on Justinian, **197**

Diptychs, their meaning and force, 83

Ennodius, St., **bishop of Pavia, asserts that God has reserved to Himself all judgment upon the successors of St. Peter, 142; his character of Acacius,** 93

Euphemius in 490 succeeds Fravita at Constantinople, 96; opposes the emperor Anastasius, but signs his Henotikon, 97; begs for reconciliation with Pope Felix, but will not give up Acacius, **97**; recognises the authority of Pope Gelasius, 103-5; deposed by the emperor through the Resident Council in 496, 114

Eutychius, patriarch of Constantinople, 239; presides over the Fifth Council, 240; consecrates Santa Sophia in 563, 244; is deposed by Justinian in 565, 245

Felix III., Pope, 483-492, 71; his letter to **the** emperor Zeno, stating his suc**cession** from St. Peter, **72**; his letter **to Acacius, 73**; holds a council in 484 and deposes Acacius, 75; his **sentence**, recounting the misdeeds of Acacius, 76-8; the synodal sentence signed by the Pope alone, which is justified by the Roman synod, 79; denounces Acacius to the emperor Zeno, 80; his utter helplessness as **to** secular support when he thus writes, 82, 88; writes afresh to the emperor Zeno that the Apostle Peter speaks in him as his Vicar, 94; delays to grant communion to Fravita, **successor** of Acacius, 94; dies after **nine** years of pontificate, 97

Filicaja, quoted, 91

Franks, made great by the Catholic faith, 44, 348; so found a kingdom, while Ostrogoths **and** Visigoths lose it, 348

Fravita, **succeeds** Acacius at Constantinople, **and** begs for the Pope's recognition, **93**; dies after three months, 96

Gelasius, Pope, 492, 98; condition of the Empire and Church at his accession, 98-9; writes to Euphemius, who will cede everything except the person of Acacius, 103-5; the bishops of Eastern Illyricum profess their obedience to the Apostolic See, **105-6**; to whom the Pope declares that the see of Constantinople has no precedence over other bishops, 107; that the Holy See, in virtue of its Principate, confirms every council, 109; his great letter to the emperor Anastasius defines the domain of the Two Powers, 110; the Primacy instituted by Christ, acknowledged by the Church, 111; in the Roman synod of 496, declares the divine Primacy of the Roman See, the second rank of Alexandria, and the third of Antioch, as sees of Peter, 113; the three Councils of Nicæa, Ephesus in 431, and Chalcedon, to be general, 116; omits the Council of Constantinople in 381, 116; death of Gelasius, and character of the time of his sitting, 118; calls Odoacer "barbarian and heretic," 68

Gennadius, bishop of Constantinople, 458-71, 64

Gibbon, acknowledges the two great victories of the Church, 325; and the work of the Church in the Spanish monarchy, 322; and the influence of bishops in establishing the French monarchy, 329

Glycerius, Roman emperor, 21

Gregorovius, "Geschichte der Stadt Rom.," quoted, 9, 11, 13, 14, 23, 42, 208, 222, 245, 247, 272-3, 275

Gregory, St., the Great, his ancestry, 276; state of Rome described by his predecessor Pope Pelagius, 277; elected Pope, 590—tries for six months to escape, 278; describes the work he was undertaking, 279; and the misery of Rome in the words of Ezechiel, 281; the Rome of St. Leo and the Rome of St. Gregory, 284; his works done out of this Rome, 285-7; the Lombard descent on Italy, 288; alludes to a strange occurrence in St. Agatha dei Goti, 21; refers to his great-grandfather, Pope Felix III., 81; describes St. Benedict rebuking Totila, 215; his right of reporting injustice to the emperor, 260; his Primacy untouched by Rome's calamities, 292; describes his Primacy to the empress Constantina, 295; identifies to her his authority with that of St. Peter, 296; also to the emperor Mauritius, 299, and to the Lombard queen Theodelinda, 312; and to the king of the Franks, 312; and to Rechared, Gothic king of Spain, 319; and in the appointment of the English hierarchy, 315; his inference from the original patriarchal sees being all sees of Peter, 301; exposes the contrast between the assumed title of the patriarch of Constantinople and his own Principate, 302-7; his title, "Servant of the servants of God," expresses his administration, 308; as fourth Doctor of the western Church, 334; as chief artificer in the Church's second victory, 335; England indebted to him, both for hierarchy and civil constitution, 336; his action as bishop, metropolitan, patriarch, and Pope, 337; councils held by him at Rome, 338; defends the liberties of monasteries against bishops, 339; and as metropolitan succours distressed bishoprics, 340; called the father of the monks, 341; compared with St. Leo in the exercise of the Primacy, 342; continues the struggle of the Popes from St. Sylvester to maintain the Nicene constitution, 350

Gregory of Tours, St., notes the prospering of the Catholic, and the decline of the Arian kingdoms, 123; attests St. Gregory's flight from the papacy, 279

Guizot, his witness to the action of the hierarchy, 54

Hefele, "Concilien-geschichte," quoted, 93, 100, 114, 116, 128, 136, 137, 139, 142, 202, 232

Hergenröther, Card., quoted, "Kirchengeschichte," 26, 114, 185, 232, 244; "Photius, sein Leben," 46, 47, 68, 75, 78, 83, 92, 93, 104, 128, 129, 143, 159, 165, 170, 187, 196, 203, 205, 207, 228, 230, 232, 245, 270, 271

Hilarus, Pope, 16

Hormisdas, deacon, elected Pope in 514, 149; sends a legation to the emperor Anastasius, who had applied to his fatherly affection, 150; instruction given to his legates, 151-8; orders them not to be introduced by the bishop of Constantinople, 157; conditions of reunion proposed by him to the emperor, 158; is deceived by the emperor, and denounces the treachery of Greek diplomacy, 160; is appealed to by the Syrian Archimandrites, 161; resolves how to terminate the Acacian schism, 164; his formulary of union accepted by the East, 167; dies in 523, 193

Hurter's "Geschichte Papst Innocenz des Dritten," the papal idea carried out through generations, 353-5

Ignatius, St., of Antioch, quoted, 12

Jerome, St., the result which he did not foresee, 57

John, patriarch of Constantinople, accepts the formulary of Pope Hormisdas, 166
John I., Pope, **martyred by Theodorick**, 193
John II., Pope, praises Justinian for acknowledging the Primacy, and confirms his confession of faith, 191
John Talaia, elected patriarch of Alexandria, 68; offends Acacius, 69; flies for refuge to Pope Simplicius, 71; is supported by Pope Felix, **75**; made bishop of Nola by Pope Felix, **92**
John the Faster, patriarch of Constantinople, assumes a **scandalous** title, 299; holds **to Gregory the** position of Anatolius to Leo, 307
Justin I., made emperor, 162; writes to Pope Hormisdas, 163; announces to him the condemnation of Acacius, 169; his reign of nine years, 198
Justinian, his origin, 162; entreats Pope Hormisdas to restore unity, 164; acknowledges to Pope John II. his Primacy, 189; enacts the *Pandects*, 192; acknowledged the Pope's Primacy all his life, 195; his character as legislator, 197; recovers North Africa, 199; begins the Gothic war, 206; domineers over the eastern Church, 227-32; acknowledges the dignity of Pope Vigilius, 232; persecutes him, 232-40; issues dogmatic decrees, 236, 242; issues Pragmatic Sanction for Italy, 243; deposes his patriarch Eutychius, 244; his conception of Church and State, 248-56; makes bishops and governors exercise mutual supervision, 257; completeness and cordiality of his alliance with the Church, 261; his spirit the opposite to that of modern governments, 262; how far he maintains, how far goes beyond, the imperial idea, 264-9; result spiritual and temporal of his reign, 270

Kurth, quoted "Les Origines de la Civilisation modern," 41; on the policy of Justinian, 255; the Church's power over the new nations, **333**

Leander, St., archbishop of Seville, becomes an intimate friend of St. Gregory during his nunciature at Constantinople, 277; receives the pallium from St. Gregory, 317, 321

Leo I., St., his universal Pastorship acknowledged by the Church in General Council, 1-3; and the succession of the Popes during 400 years, from St. Peter, 3; rescues Rome from Attila, and from Genseric, 7-8; his character, acts, and times, 15; stands between the two great victories of the Church, and represents both, 25-6; the result which **St. Leo** did not foresee, 57; his prescience of usurpation from the Byzantine bishop, 60; his prescience of what the bishops of Constantinople aimed at, **307**; draws out the office and functions of the nuncio, 338
Leo I., emperor, 467, 62; dies in 474, 63
Leo II., an infant, succeeds for a few months, 63
Liberatus, "Breviarium," quoted, 208, 209
Libius Severus, Roman emperor, 16
Lombards, their descent on Italy and uncivilised savagery, 287-91; **for ever strive** to possess Rome, **but never succeed, 347**

Macedonius, bishop **of** Constantinople, feels **his** unlawful appointment, 143; persecuted during fifteen years, and finally deposed **by the** emperor Anastasius, 144-8; refuses to give up the Council of Chalcedon, but will not surrender the memory of Acacius, and never enjoys communion with the Pope, 144-8
Majorian, Roman emperor, 14
Martyrdom, Papal, of 300 years, 10, 54
Mausoleum of Hadrian, stripped of its statues, 211; an apparition of St. Michael changes its name, 278
Mennas, patriach of Constantinople, 228-239

Nepos, Roman emperor, 21

Odoacer, extinguishes the western emperor, 22; named Patricius of the Romans by the emperor Zeno, 35; slain by Theodorick, 38; his exaltation foretold by St. Severinus, 22
Olybrius, Roman emperor, 20
Orosius, an important anecdote preserved by him, 43

Pallium, sent by the Pope to the chief bishop in each province, 337; the duties and powers which it carried with it, 337

INDEX.

Papal election, the freedom of, assailed by Odoacer, 194, 292; by Theodorick and Justinian, 210, 292
Pelagius II., Pope, 578-590, describes the state of Rome, 277
Petra Apostolica in the sixty Popes preceding Gregory, 352; in the Popes from St. Gregory to Innocent III., 353; in the Popes from Innocent III. to Leo XIII., 355; sustained by opposing forces, 359
Philips, "Kirchenrecht," his judgment of Theodorick, 41; on Byzantine succession, 61
Primacy, the Roman, its denial suicidal in all who believe one holy Catholic Church, 3-4; the creator of Christendom, 5, 6, 10, 57-8; tested by the division of the empire, 51; still more by the extinction of the western emperor, 53; witness to it by Guizot, 55; saves, in the seven successors of St. Leo, the eastern Church from becoming Eutychean, 179-86; developed by the sufferings of sixty years, 188; acknowledged by the Council of Africa after the expulsion of the Vandals, 201; defined by the Vatican Council, as held by St. Gregory I., 307; saves the western bishops from absorption in their several countries, 330; preserver of civil liberties, 333; resister of Byzantine despotism, 333; its development from St. Leo I. to St. Gregory I., 342; confirmed and illustrated by civil disasters, 346; as Rome, the secular city, diminishes, the Primacy advances, 357

Rechared, king of the Spanish Visigoths, converted, 318; his letter to St. Gregory informing him of his conversion, 321
Reumont, "Geschichte der Stadt Rom.," quoted, over-lordship of Byzantium, 19; Odoacer, Patricius at Rome, 35; picture of Theodorick, 36; of his government, 38; sparing of St. Peter's and St. Paul's, 213; Totila's deeds, 215; Narses made Patricius of Rome, 245; the Pragmatic Sanction, 246
Riffel, "Kirche und Staat," quoted, 190, 251, 253, 254, 255, 256, 267.
Röhrbacher, the German edition of the history, quoted, 128, 142, 162, 192, 198, 199, 200, 202, 205, 245, 303, 305

Rome, its fall as a city coeval with the universal recognition of the Papal Primacy, 6-10; this fall and this recognition traced from Constantine to St. Gregory, 356-8; imperial, its death agony of twenty-one years, 23; its sufferings in the Gothic war, 210-23; the new city, from Narses, lives only by the Primacy, 294; its extreme misery in the days of St. Gregory, 281, 284
Romulus Augustulus, Roman emperor, 21

Saxons, rudest of Teuton tribes, humanised by St. Gregory, 348
Sidonius Apollinaris, picture of the Roman senate, 17; description of Rome in 467, 18; makes Rome acknowledge the over-lordship of the East, 19; describes the Roman baths, 19
Silverius, St., Pope, elected in 536, 205; deposed by Belisarius, at the instigation of Theodora, 208; martyred in the island of Patmaria, 209
Simplicius, Pope, his outlook from Rome, 45; his letter to the emperor Zeno, 66
Symmachus, elected Pope in 498, 128; his letter to the eastern emperor, 129; compares the imperial and the papal power, 131; they are the two heads of human society, 133; Catholic princes acknowledge Popes on their accession, 134; inferences to be deduced from this letter, 136; the Synodus Palmaris refuses to judge the Pope, 136; addressed by eastern bishops in their misery as a father by his children, 149; dies in 514, 149

Theodora, empress, her promises to Vigilius, 208; her violent deposition of Pope Silverius, 209
Theodorick, the Ostrogoth, how nurtured, 36; marches on Italy, 37; which he conquers, and slays Odoacer, 38; character of his reign, 39; slays Pope John I., and his own ministers, Boethius and Symmachus, 41, 329; judgment of him by St. Gregory, 41; contrast with Clovis, 42; his kingdom came to nothing, 43; asks the title of king from the emperor Anastasius, 128; determines the election of Pope Symmachus against Laurentius, 129; induced

to send a bishop as visitor of the Roman Church, 137; said by the emperor to have the charge of governing the Romans committed to him, 159; his ability and family connections, 177; final failure of his state, his family, and people, 328-9; his attempt to maintain Arianism in the West foiled, 347

Thierry, "Derniers temps de l'Empire d'Occident," 20

Tillemont, quoted, 64

Totila, elected Gothic king, **214**; is warned by St. Benedict, 215; takes Rome, 216; takes Rome, its fourth capture, 218; killed at Taginas, **219**

Valens, emperor, poisons the western empire with Arianism, 50, 92

Valentinian III. his edict in 447 terms the Pope, Leo I., *principem episcopalis coronæ*, 56; murdered by Maximus, 13

Vere, A. de, quoted, "Legends and Records," 1, 12; "Chains of St. Peter," 272

Vigilius, made Pope by Belisarius, **209**; summoned to Constantinople by Justinian, 226; his persecution there, 232-243; his dignity as Pope left unimpaired, 293

Vitiges besieges Rome, and ruins the aqueducts and Campagna, 210-13; carried a captive to Constantinople, 214

Wandering of the nations, **26-35**

Zeno, eastern emperor, 63; second of the theologising emperors, 47; his conduct and character, 63; matched with the emperor Valens, 92; his death, **91, 99**

SELECTION

FROM

BURNS & OATES'

Catalogue

OF

PUBLICATIONS.

LONDON: BURNS AND OATES, Ld.

ORCHARD ST., W., & 63 PATERNOSTER ROW, E.C.

NEW YORK: 9 BARCLAY STREET.

1888.

NEW BOOKS.

The Holy See and the Wandering of the Nations. By T W. ALLIES, K.C.S.G. Demy 8vo, cloth, 10s. 6d.

Characteristics from the Writings of Archbishop Ullathorne, together with a Bibliographical account of the Archbishop's Work's By the Rev. M. F. GLANCEY, of St. Mary's College, Oscott. [Immediately.

St. Peter, Bishop of Rome; or, the Roman Episcopate of the Prince of the Apostles, proved from the Fathers, History and Chronology, and illustrated by arguments from other sources. By the Rev. T. LIVIUS, C.SS.R., M.A., Oriel College, Oxford. Dedicated to his Eminence Cardinal Newman. Demy 8vo, 12s.

A Menology of England and Wales; or, Brief Memorials of the British and English Saints, arranged according to the Calendar. Together with the Martyrs of the 16th and 17th centuries. Compiled by order of the Cardinal Archbishop and the Bishops of the Province of Westminster, by the Rev. RICHARD M. STANTON, Priest of the Oratory. In one volume. Demy 8vo, cloth, 14s.

The Canons and Decrees of the Sacred and Œcumenical Council of Trent, celebrated under the Sovereign Pontiffs, Paul III., Julius III., and Pius IV., translated by the Rev. J. WATERWORTH. To which are prefixed Essays on the External and Internal History of the Council. A new edition. Demy, 8vo, cloth, 10s. 6d.

Records of the English Catholics of 1715. Edited by JOHN ORLEBAR PAYNE, M.A. With a complete index. Demy 8vo, Cloth gilt, 15s.

Explanation of the Psalms and Canticles in the Divine Office. By ST. ALPHONSUS LIGUORI. Translated from the Italian by THOMAS LIVIUS, C.SS.R. With a Preface by his Eminence Cardinal MANNING. Crown 8vo, cloth, 7s. 6d.

Life of St. Anastasia. Translated by Miss MARGARET HOWITT, and edited by the Rev. KENELM VAUGHAN. [Immediately.

Lives of the Saints and Blessed of the Three Orders of St. Francis. Translated from the original "Auréole Séraphique" of the Very Rev. FATHER LEON, ex-Provincial of the Friars Minor of the Observance. Complete in 4 vols., price 8s. 6d. each.

Life of St. Patrick, Apostle of Ireland. By the Rev. W. B. MORRIS, of the Oratory. A new, revised, and greatly enlarged edition. Cloth, 5s.

"Jesu's Psalter"; What it was at its origin, and as consecrated by the use of many Martyrs and Confessors. With Chant for its more solemn recitation. By the Rev. SAMUEL HEYDON SOLE, Priest of Chipping Norton. Cloth gilt, 3s. 6d.

St. Peter's Chains; or, Rome and the Italian Revolution. A series of Sonnets. By AUBERY DE VERE. Cloth, 2s.

SELECTION
FROM
BURNS AND OATES' CATALOGUE OF PUBLICATIONS.

ALLIES, T. W. (K.C.S.G.)

	£ s. d.
See of St. Peter.	0 4 6
Formation of Christendom. Vols. I., II., III. each	0 12 0
Church and State as seen in the Formation of Christendom, 8vo, pp. 472, cloth	0 14 0
The Throne of the Fisherman, built by the Carpenter's Son, the Root, the Bond, and the Crown of Christendom. Demy 8vo	0 10 6

"It would be quite superfluous at this hour of the day to recommend Mr. Allies' writings to English Catholics. Those of our readers who remember the article on his writings in the *Katholik*, know that he is esteemed in Germany as one of our foremost writers."—*Dublin Review.*

ALLIES, MARY.

Leaves from St. Augustine. With preface by T. W. Allies, K.C.S.G. Crown 8vo	0 6 0

"The plain, outspoken, yet truly Christian doctrine of the great Bishop of Hippo has an honest hearty ring about it which contrasts strangely with the weak-kneed theology of those who would cut and trim the Gospel to the taste of worldly society."—*Morning Post.*

"Welcome to such volumes, and were there many of them."—*Weekly Register.*

ALLNATT, C. F. B.

Cathedra Petri. Third and Enlarged Edition. Paper.	0 5 0

"Invaluable to the controversialist and the theologian, and most useful for educated men inquiring after truth or anxious to know the positive testimony of Christian antiquity in favour of Papal claims."—*Month.*

Which is the True Church? New Edition	0 1 4
The Church and the Sects	0 1 0

ANNUS SANCTUS:

Hymns of the Church for the Ecclesiastical Year. Translated from the Sacred Offices by various Authors, with Modern, Original, and other Hymns, and an Appendix of Earlier Versions. Selected and Arranged by ORBY SHIPLEY, M.A. In stiff boards.	0 3 6
Plain Cloth, lettered	0 5 0
Edition de luxe	0 10 6

ANSWERS TO ATHEISTS: OR NOTES ON
Ingersoll. By the Rev. A Lambert, (over 100,000 copies sold in America). Ninth edition. Paper . . . £0 0 6
Cloth . . . 0 1 0

B. N.
The Jesuits: their Foundation and History. 2 vols. crown 8vo, cloth, red edges . . . 0 15 0
"The book is just what it professes to be—*a popular history*, drawn from well-known sources," &c.—*Month.*

BACQUEZ, L'ABBE.
The "Divine Office": From the French of l'Abbé Bacquez, of the Seminary of St. Sulpice, Paris. Edited by the Rev. Father Taunton, of the Congregation of the Oblates of St. Charles. Cloth 0 6 0
"The translation of this most edifying work from the walls of St. Sulpice, the source of so much sacerdotal perfection, comes to us most opportunely, and we heartily commend it to the use of the clergy and of the faithful." THE CARDINAL ARCHBISHOP OF WESTMINSTER.
"A very complete manual, learned, wholesome, and devout."—*Saturday Review.*

BELLECIO, FATHER ALOYSIUS, (S.J.).
Spiritual Exercises, according to the Method of St. Ignatius of Loyola. Translated from the Italian Version of Father Anthony Bresciani, S.J., by William Hutch, D.D. Third edition . . . 0 2 6

BORROMEO, LIFE OF ST. CHARLES.
From the Italian of Peter Guissano. 2 vols. . . 0 15 0
"A standard work, which has stood the test of succeeding ages; it is certainly the finest work on St. Charles in an English dress."—*Tablet.*

BOWDEN, REV. H. S. (of the Oratory) Edited by
Dante's Divina Commedia: Its scope and value. From the German of FRANCIS HETTINGER, D.D. With an engraving of Dante. Crown 8vo . . 0 10 6
"All that Venturi attempted to do has been now approached with far greater power and learning by Dr. Hettinger, who, as the author of the 'Apologie des Christenthums,' and as a great Catholic theologian, is eminently well qualified for the task he has undertaken."—*The Saturday Review.*

BRIDGETT, REV. T. E. (C.SS.R.).
Discipline of Drink . . . 0 3 6
"The historical information with which the book abounds gives evidence of deep research and patient study, and imparts a permanent interest to the volume, which will elevate it to a position of authority and importance enjoyed by few of its compeers."—*The Arrow.*

Our Lady's Dowry; how England Won and Lost that Title. Second Edition . . . 0 9 0
"This book is the ablest vindication of Catholic devotion to Our Lady, drawn from tradition, that we know of in the English language."—*Tablet.*

BRIDGETT, REV. T. E. (C.SS.R.)—*continued*.

Ritual of the New Testament. An essay on the principles and origin of Catholic Ritual in reference to the New Testament. Third edition . . . £0 5 0

The Life of the Blessed John Fisher. With a reproduction of the famous portrait of Blessed JOHN FISHER by HOLBEIN, and other Illustrations. Cloth . 0 7 6

BRIDGETT, REV. T. E. (C.SS.R.), Edited by.

Suppliant of the Holy Ghost: a Paraphrase of the 'Veni Sancte Spiritus.' Now first printed from a MS. of the seventeenth century composed by Rev. R. Johnson, with other unpublished treatises by the same author. Second edition. Cloth . . . 0 1 6

Souls Departed. By CARDINAL ALLEN. First published in 1565, now edited in modern spelling by the Rev. T. E. Bridgett 0 6 0

CASWALL, FATHER.

Catholic Latin Instructor in the Principal Church Offices and Devotions, for the Use of Choirs, Convents, and Mission Schools, and for Self-Teaching. 1 vol., complete 0 3 6

Or Part I., containing Benediction, Mass, Serving a Mass, and various Latin Prayers in ordinary use . 0 1 6

May Pageant: A Tale of Tintern. (A Poem) Second edition 0 2 0

Poems 0 5 0

Lyra Catholica, containing all the Breviary and Missal Hymns, with others from various sources. 32mo, cloth, red edges 0 2 6

CATHOLIC BELIEF: OR, A SHORT AND

Simple Exposition of Catholic Doctrine. By the Very Rev. Joseph Faà di Bruno, D.D. Sixth edition. . . . . . . Price 6d.; post free, 0 0 8½

Cloth, lettered, ... 0 0 10

Also an edition on better paper and bound in cloth, with gilt lettering and steel frontispiece 0 2 0

CHALLONER, BISHOP.

Meditations for every day in the year. New edition. Revised and edited by the Right Rev. John Virtue, D.D., Bishop of Portsmouth. 8vo. 5th edition . 0 3 0

And in other bindings.

COLERIDGE, REV. H. J. (S.J.)

(See Quarterly Series.)

DEHARBE, FATHER JOSEPH, (S.J.)

A History of Religion, or the Evidences or the Divinity of the Christian Religion, as furnished by its History from the Creation of the World to our own Times. Designed as a Help to Catechetical Instruction in Schools and Churches. Pp. 628. . . . . . net £0 8 6

DEVAS, C. S.

Studies of Family Life: a contribution to Social Science. Crown 8vo. 0 5 0

"We recommend these pages and the remarkable evidence brought together in them to the careful attention of all who are interested in the well-being of our common humanity."—*Guardian*
"Both thoughtful and stimulating."—*Saturday Review.*

DRANE, AUGUSTA THEODOSIA.

History of St. Catherine of Siena and her Companions. A new edition in two vols. 0 12 6

"It has been reserved for the author of the present work to give us a complete biography of St. Catherine. . . . Perhaps the greatest success of the writer is the way in which she has contrived to make the Saint herself live in the pages of the book."—*Tablet.*

DUKE, REV. H. C.

King, Prophet, and Priest: or, a Course of Lectures on the Catholic Church. Cloth . . . 0 6 6

"Seventeen admirable lectures full of instruction, learned as well as simple . . . singularly well arranged and very clearly expressed."—*Tablet.*

ENGLISH CATHOLIC NON-JURORS OF 1715.

Being a Summary of the Register of their Estates, with Genealogical and other Notes, and an Appendix of Unpublished Documents in the Public Record Office. Edited by the late Very Rev. E. E. Estcourt, M.A., F.S.A., Canon of St. Chad's, Birmingham, and John Orlebar Payne, M.A. 1 vol., demy 8vo. . 1 1 0

"This handsomely printed volume lies before us. Every student of the history of our nation, or of families which compose it, cannot but be grateful for a catalogue such as we have here."—*Dublin Review.*
"Most carefully and creditably brought out. . . . From first to last full of social interest, and it contains biographical details for which we may search in vain elsewhere."—*Antiquarian Magazine.*

EYRE, MOST REV. CHARLES, (Abp. of Glasgow).

The History of St. Cuthbert; or, An Account of his Life, Decease, and Miracles. Third Edition. Illustrated with maps, charts, &c., and handsomely bound in cloth. Royal 8vo 0 14 0

"A handsome, well appointed volume, in every way worthy of its illustrious subject. . . . The chief impression of the whole is the picture of a great and good man drawn by a sympathetic hand."—*Spectator.*

FABER, REV. FATHER.

	£	s	d
All for Jesus	0	5	0
Bethlehem	0	7	0
Blessed Sacrament	0	7	6
Creator and Creature	0	6	0
Ethel's Book of the Angels	0	5	0
Foot of the Cross	0	6	0
Growth in Holiness	0	6	0
Hymns	0	6	0
Notes on Doctrinal and Spiritual Subjects, 2 vols. each	0	5	0
Poems	0	5	0
Precious Blood	0	5	0
Sir Lancelot	0	5	0
Spiritual Conferences	0	6	0
Life and Letters of Frederick William Faber, D.D., Priest of the Oratory of St. Philip Neri. By John Edward Bowden of the same Congregation	0	6	0

FOLEY, HENRY (S.J.)

	£	s	d
Records of the English Province of the Society of Jesus. Vol. I., Series I. Demy 8vo, 720 pp. net	1	6	0
Vol. II., Series II., III., IV. Demy 8vo, 622 pp. net	1	6	0
Vol. III., Series V., VI., VII., VIII. Demy 8vo, over 850 pp.	1	10	0
Vol. IV. Series IX., X., XI. Demy 8vo, 750 pp. net	1	6	0
Vol. V., Series XII. Demy 8vo, nearly 1100 pp., with nine Photographs of Martyrs . net	1	10	0
Vol. VI., Diary and Pilgrim-Book of the English College, Rome. The Diary from 1579 to 1773, with Biographical and Historical Notes. The Pilgrim-Book of the Ancient English Hospice attached to the College from 1580 to 1656, with Historical Notes. Demy 8vo, pp. 796 net	1	6	0
Vol. VII. Part the First: General Statistics of the Province; and Collectanea, giving Biographical Notices of its Members and of many Irish and Scotch Jesuits. With 20 Photographs net	1	6	0
Vol. VII. Part the Second: Collectanea, Completed; With Appendices. Catalogues of Assumed and Real Names: Annual Letters; Biographies and Miscellanea. net	1	6	0

"As a biographical dictionary of English Jesuits, it deserves a place in every well-selected library, and, as a collection of marvellous occurrences, persecutions, martyrdoms, and evidences of the results of faith, amongst the books of all who belong to the Catholic Church."—*Genealogist*.

FORMBY, REV. HENRY.

	£	s	d
Monotheism: in the main derived from the Hebrew nation and the Law of Moses. The Primitive Religion of the City of Rome. An historical Investigation. Demy 8vo.	0	5	0

FRANCIS DE SALES, ST.: THE WORKS OF.

Translated into the English Language by the Rev. H. B. Mackey, O.S.B., under the direction of the Right Rev. Bishop Hedley, O.S.B.

Vol. I. Letters to Persons in the World. Cloth . £0 6 0

"The letters must be read in order to comprehend the charm and sweetness of their style."—*Tablet.*

Vol. II.—The Treatise on the Love of God. Father Carr's translation of 1630 has been taken as a basis, but it has been modernized and thoroughly revised and corrected. 0 9 0

"To those who are seeking perfection by the path of contemplation this volume will be an armoury of help."—*Saturday Review.*

Vol. III. The Catholic Controversy. . . 0 6 0

"No one who has not read it can conceive how clear, how convincing, and how well adapted to our present needs are these controversial 'leaves.'"—*Tablet.*

Vol. IV. Letters to Persons in Religion. [Just out. 0 6 0

*** Other vols. in preparation.

Devout Life	0	1 6
Manual of Practical Piety	0	3 6
Spiritual Combat. Pocket size, 32mo, cloth .	0	1 0

GALLWEY, REV. PETER (S.J.)

Precious Pearl of Hope in the Mercy of God, The. Translated from the Italian. With Preface by the Rev. Father Gallwey. Cloth 0 4 6

Lectures on Ritualism and on the Anglican Orders. 2 vols. 0 8 0

Or may be had separately.

GIBSON, REV. H.

Catechism Made Easy. Being an Explanation of the Christian Doctrine. 2 vols., cloth . . . 0 7 6

"This work must be of priceless worth to any who are engaged in any form of catechetical instruction. It is the best book of the kind that we have seen in English."—*Irish Monthly.*

GILLOW, JOSEPH.

Literary and Biographical History, or, Bibliographical Dictionary of the English Catholics. From the Breach with Rome, in 1534, to the Present Time. Vols. I., II. and III. *cloth, demy 8vo* . *each.* 0 15 0

"The patient research of Mr. Gillow, his conscientious record of minute particulars, and especially his exhaustive bibliographical information in connection with each name, are beyond praise."—*British Quarterly Review.*

"No such important or novel contribution has been made to English bibliography for a long time."—*Scotsman.*

The Haydock Papers. Illustrated. Demy 8vo. . 0 7 6

HEDLEY, BISHOP.

Our Divine Saviour, and other Discourses. Crown 8vo. £0 6 0

"A distinct and noteworthy feature of these sermons is, we certainly think, their freshness—freshness of thought, treatment, and style; nowhere do we meet pulpit commonplace or hackneyed phrase—everywhere, on the contrary, it is the heart of the preacher pouring out to his flock his own deep convictions, enforcing them from the 'Treasures, old and new,' of a cultivated mind.'"—*Dublin Review.*

HERGENROTHER, DR.

Catholic Church and Christian State. On the Relation of the Church to the Civil Power. From the German. 2 vols., paper 1 0 0

HUMPHREY, REV. W. (S.J.)

The Divine Teacher: A Letter to a Friend. With a Preface in Reply to No. 3 of the English Church Defence Tracts, entitled "Papal Infallibility."
Fifth edition. Cloth 0 2 6
Sixth edition. Wrapper 0 1 0
Mary Magnifying God. May Sermons. Fifth edition 0 2 6
Other Gospels; or, Lectures on St. Paul's Epistle to the Galatians. Crown 8vo, cloth . . . 0 4 0
The Written Word; or, Considerations on the Sacred Scriptures 0 5 0
Mr. Fitzjames Stephen and Cardinal Bellarmine . 0 1 0
Suarez on the Religious State: A Digest of the Doctrine contained in his Treatise, "De Statû Religionis." 3 vols., pp. 1200. Cloth, roy. 8vo. . 1 10 0

"This laborious and skilfully executed work is a distinct addition to English theological literature. Father Humphrey's style is quiet, methodical, precise, and as clear as the subject admits. Every one will be struck with the air of legal exposition which pervades the book. He takes a grip of his author, under which the text yields up every atom of its meaning and force."—*Dublin Review.*

LEE, REV. F. G. (D.D.)

Edward the Sixth: Supreme Head. Crown 8vo . 0 10 6

"In vivid interest and in literary power, no less than in solid historical value, Dr. Lee's present work comes fully up to the standard of its predecessors; and to say that is to bestow high praise. The book evinces Dr. Lee's customary diligence of research in amassing facts, and his rare artistic power in welding them into a harmonious and effective whole."—*John Bull.*

LIFE OF FATHER CHAMPAGNAT,

Founder of the Society of the Little Brothers of Mary. Containing a portrait of Fr. CHAMPAGNAT, and four full page illustrations. Demy 8vo . . . 0 8 0

"A work of great practical utility, and one eminently suited to these times."—*Tablet.*

"A serious and able essay on the science and art of the Christian education of children, exemplified in the career of one who gave his life to it."—*Dublin Review.*

LIGUORI, ST. ALPHONSUS.

New and Improved Translation of the Complete Works of St. Alphonsus, edited by the late Bishop Coffin:—

Vol. 1. The Christian Virtues, and the Means for Obtaining them. Cloth elegant £0 4 0

Or separately:—
1. The Love of our Lord Jesus Christ . . . 0 1 4
2. Treatise on Prayer. *(In the ordinary editions a great part of this work is omitted)* . . . 0 1 4
3. A Christian's rule of Life 0 1 0

Vol. II. The Mysteries of the Faith—The Incarnation; containing Meditations and Devotions on the Birth and Infancy of Jesus Christ, &c., suited for Advent and Christmas 0 3 6
   Cheap edition 0 2 0

Vol. III. The Mysteries of the Faith—The Blessed Sacrament 0 3 6
   Cheap edition 0 2 0

Vol. IV. Eternal Truths—Preparation for Death . 0 3 6
   Cheap edition 0 2 0

Vol. V. Treatises on the Passion, containing "Jesus hath loved us," &c. 0 3 0
   Cheap edition 0 2 0

Vol. VI. Glories of Mary. New edition . . 0 3 6
   With Frontispiece, cloth 0 4 6
   Also in better bindings.

MANNING CARDINAL.

Blessed Sacrament the Centre of Immutable Truth. A new revised edition. 0 1 0
Confidence in God. Fourth edition 0 1 0
England and Christendom 0 10 6
Eternal Priesthood. Seventh Edition . . . 0 2 6
Four Great Evils of the Day. Fifth Edition. Paper 0 2 6
   Cloth 0 3 6
Fourfold Sovereignty of God. Third edition Paper 0 2 6
   Cloth 0 3 6
Glories of the Sacred Heart. Fourth edition. . . 0 6 0
Grounds of Faith. Seventh edition. 0 1 6
Holy Gospel of our Lord Jesus Christ according to St. John. With a Preface by His Eminence. . . 0 1 0
Religio Viatoris. Third Edition. Wrapper. . . 0 1 0
   Cloth. 0 2 0
Independence of the Holy See. Second Edition. . 0 5 0
Internal Mission of the Holy Ghost. Fourth edition . 0 8 6
Love of Jesus to Penitents. Seventh edition . . 0 1 6
Miscellanies. 2 vols. (Vol. III. is in preparation.) .. 0 15 0
Office of the Holy Ghost under the Gospel . . 0 1 0
Petri Privilegium 0 10 6
Praise, A Sermon on; with an Indulgenced Devotion. 0 1 0
Sermons on Ecclesiastical Subjects. Vols. I. II. and III. . . . . . . each 0 6 0

MANNING, CARDINAL—*continued*.

Sin and its Consequences. Sixth edition	£0	6	0
Temporal Mission of the Holy Ghost. Third edition	0	8	6
Temporal Power of the Pope. Third edition	0	5	0
The Office of the Church in Higher Education	0	0	6
True Story of the Vatican Council. Second Edition.	0	5	0

MANNING, CARDINAL, Edited by.

Life of the Curé of Ars. New edition, enlarged.	0	4	0

MIVART, PROF. ST. GEORGE (M.D., F.R.S.)

Nature and Thought. Second edition	0	4	0

"The complete command of the subject, the wide grasp, the subtlety, the readiness of illustration, the grace of style, contrive to render this **one** of the **most** admirable books of its class."—*British Quarterly Review.*

A Philosophical Catechism. Fifth edition	0	1	0

"It should become the *vade mecum* of Catholic students."—*Tablet.*

MORRIS, REV. JOHN (S.J.)

Letter Books of Sir Amias Poulet, keeper of Mary Queen of Scots. Demy 8vo	0	10	6
Troubles of our Catholic Forefathers, related by themselves. Second Series. 8vo, cloth.	0	14	0
Third Series	0	14	0
The Life of Father John Gerard, S.J. Third edition, rewritten and enlarged	0	14	0
The Life and Martyrdom of St. Thomas Becket. Second and enlarged edition. In one volume, large post 8vo, cloth, pp. xxxvi., 632,	0	12	6
or bound in two parts, cloth	0	13	0

MURPHY, J. N.

Chair of Peter. Third edition, with the statistics, &c., brought down to the present day. 720 pages. Crown 8vo	0	6	0

"In a series of clearly written chapters, precise in statement, excellently temperate in tone, the author deals with just those questions regarding the power, claims, and history of the Roman Pontiff which are at the present time of most actual interest."—*Dublin Review.*

NEWMAN, CARDINAL.

Annotated Translation of Athanasius. 2 vols. each	0	7	6
Apologia pro Vitâ suâ	0	6	0
Arians of the Fourth Century, The	0	6	0
Callista. An Historical Tale.	0	5	6
Difficulties of Anglicans. Two volumes—			
Vol. I. Twelve Lectures.	0	7	6
Vol. II. Letter to Dr. Pusey and to the Duke of Norfolk	0	5	6
Discussions and Arguments	0	6	0
Doctrine of Justification	0	5	0

NEWMAN, CARDINAL—*continued*.

	£	s	d
Dream of Gerontius.	0	0	6
Essay on Assent	0	7	6
Essay on the Development of Christian Doctrine	0	6	0
Essays Critical and Historical. Two volumes, with Notes . . . . . . each	0	6	0
Essays on Miracles, Two. 1. Of Scripture. 2. Of Ecclesiastical History	0	6	0
Historical Sketches. Three volumes . . each	0	6	0
Idea of a University. Lectures and Essays	0	7	0
Loss and Gain. Ninth Edition .	0	5	6
Occasional Sermons	0	6	0
Parochial and Plain Sermons. Eight volumes. each	0	5	0
Present Position of Catholics in England.	0	7	0
Sermons on Subjects of the Day.	0	5	0
Sermons to Mixed Congregations	0	6	0
Theological Tracts	0	8	0
University Sermons	0	5	0
Verses on Various Occasions.	0	5	6
Via Media. Two volumes, with Notes . each	0	6	0

NORTHCOTE, VERY REV. J. S. (D.D.)

Roma Sotterranea; or, An Account of the Roman Catacombs. New edition. Re-written and greatly enlarged. This work is in three volumes, which may at present be had separately—

	£	s	d
Vol. I. History .	1	4	0
Vol. II. Christian Art.	1	4	0
Vol. III. Epitaphs of the Catacombs	0	10	0
The Second and Third Volumes may also be had bound together in cloth	1	12	0
Visit to the Roman Catacombs: Being a popular abridgment of the larger work.	0	4	0
Mary in the Gospels .	0	3	6

POPE, THOMAS ALDER, M.A. (of the Oratory.)

	£	s	d
Life of St. Philip Neri, Apostle of Rome. From the Italian of Alfonso Capecelatro. 2 vols .	0	15	0

"No former life has given us so full a knowledge of the surroundings of St. Philip. . . . To those who have not read the original we can say, with the greatest confidence, that they will find in these two well-edited volumes a very large store of holy reading and of interesting history,"—*Dublin Review.*

QUARTERLY SERIES (Edited by the Rev. H. J. Coleridge, S.J.)

	£	s	d
Baptism of the King: Considerations on the Sacred Passion. By the Rev. H. J. Coleridge, S.J.	0	7	6
Christian Reformed in mind and Manners, The. By Benedict Rogacci, of the Society of Jesus. The Translation edited by the Rev. H. J. Coleridge, S.J	0	7	6

QUARTERLY SERIES—continued.

	£ s. d.
Chronicles of St. Antony of Padua, the "Eldest Son of St. Francis." Edited by the Rev. H. J. Coleridge, S.J.	0 3 6
Colombière, Life of the Ven. Claude de la .	0 5 0
Dialogues of St. Gregory the Great: an Old English Version. Edited by the Rev. H. J. Coleridge, S.J.	0 6 0
During the Persecution. Autobiography of Father John Gerard, S.J. Translated from the original Latin by the Rev. G. R. Kingdon, S.J.	0 5 0
English Carmelite, An. The Life of Catherine Burton, Mother Mary Xaveria of the Angels, of the English Teresian Convent at Antwerp. Collected from her own Writings, and other sources, by Father Thomas Hunter, S.J.	0 6 0
Gaston de Ségur. A Biography. Condensed from the French Memoir by the Marquis de Ségur, by F. J. M. A. Partridge	0 3 6
Gracious Life, A (1566–1618); being the Life of Madame Acarie (Blessed Mary of the Incarnation), of the Reformed Order of our Blessed Lady of Mount Carmel. By Emily Bowles.	0 6 0
History of the Sacred Passion. By Father Luis de la Palma, of the Society of Jesus. Translated from the Spanish. With Preface by the Rev. H. J. Coleridge, S.J. Third edition	0 5 0
Holy Infancy Series. By the Rev. H. J. Coleridge, S.J. Vol. I. Preparation of the Incarnation.	0 7 6
,, II. The Nine Months. Life of our Lord in the Womb	0 7 6
,, III. The Thirty Years. Our Lord's Infancy and Hidden Life	0 7 6
Hours of the Passion. Taken from the Life of Christ by Ludolph the Saxon	0 7 6
Life and Teaching of Jesus Christ, in Meditations for every Day in the Year. By P. N. Avancino, S.J. 2 vols.	0 10 6
Life and Letters of St. Francis Xavier. By the Rev. H. J. Coleridge, S.J. 2 vols.	0 10 6
Life of Anne Catherine Emmerich. By Helen Ram. With Preface by the Rev. H. J. Coleridge, S.J.	0 5 0
Life of Christopher Columbus. By the Rev. A. G. Knight, S.J.	0 6 0
Life of Henrietta d'Osseville (in Religion, Mother Ste. Marie), Foundress of the Institute of the Faithful Virgin. Arranged and edited by the Rev. John George M'Leod. S.J.	0 5 6
Life of Margaret Mostyn (Mother Margaret of Jesus), Religious of the Reformed Order of our Blessed Lady of Mount Carmel (1625–1679). By the Very Rev. Edmund Bedingfield. Edited from the Manuscripts preserved at Darlington, by the Rev. H. J. Coleridge, S.J.	0 6 0

QUARTERLY SERIES—*continued*.

	£	s	d
Life of our Life: The Harmony of the Gospel, arranged with Introductory and Explanatory Chapters, Notes and Indices. By the Rev. H. J. Coleridge, S.J. 2 vols. (out of print)	0	15	0
Life of the Blessed John Berchmans. Third edition. By the Rev. F. Goldie, S.J.	0	6	0
Life of the Blessed Peter Favre, First Companion of St. Ignatius Loyola. From the Italian of Father Boero. (Out of print).	0	6	6
Life of King Alfred the Great. By Rev. A. G. Knight, S.J. Book I. Early Promise; II. Adversity; III. Prosperity; IV. Close of Life.	0	6	0
Life of Mother Mary Teresa Ball. By Rev. H. J. Coleridge, S.J. With Portrait.	0	6	6
Life of St. Jane Frances Fremyot de Chantal. By Emily Bowles. Third Edition.	0	5	0
Life of St. Bridget of Sweden. By the late F. J. M. A. Partridge.	0	6	0
Life and Letters of St. Teresa. 3 vols. By Rev. H. J. Coleridge, S.J. . . . . each	0	7	6
Life of Mary Ward. By Mary Catherine Elizabeth Chambers, of the Institute of the Blessed Virgin. Edited by the Rev. H. J. Coleridge, S.J. 2 vols., each	0	7	6
Life of Jane Dormer, Duchess of Feria. By Henry Clifford. Transcribed from the Ancient Manuscript in the possession of the Lord Dormer, by the late Canon E. E. Estcourt, and edited by the Rev. Joseph Stevenson, S.J.	0	5	0
Mother of the King, The. By the Rev. H. J. Coleridge, S.J.	0	7	6
Mother of the Church. "Sequel to Mother of the King."	0	6	0
Of Adoration in Spirit and Truth. By the Rev. J. E. Nieremberg. S.J. Old English translation. With a Preface by the Rev. P. Gallwey, S.J. A New Edition	0	6	6
Pious Affections towards God and the Saints. Meditations for every Day in the Year, and for the Principal Festivals. From the Latin of the Ven. Nicholas Lancicius, S.J. With Preface by Archbishop George Porter, S.J.	0	7	6
Prisoners of the King, a book of thoughts on the doctrine of Purgatory. By the Rev. H. J. Coleridge, S.J. New Edition.	0	5	0
Public Life of our Lord Jesus Christ. By the Rev. H. J. Coleridge, S.J. vols 1 to 9 . . . each	0	6	6
Vols 10 and 11 . . . . . each	0	6	0
Return of the King. Discourses on the Latter Days. By the Rev. H. J. Coleridge, S.J.	0	7	6

St. Mary's Convent, Micklegate Bar, York. A

CATALOGUE OF PUBLICATIONS. 15

QUARTERLY SERIES—*continued.*

	£	s.	d.
History of the Convent. Edited by the Rev. H. J. Coleridge, S.J.	0	7	6
Story of St. Stanislaus Kostka. With Preface by the Rev. H. J. Coleridge, S.J.	0	3	6
Story of the Gospels, harmonised for meditation. By the Rev. H. J. Coleridge, S.J.	0	7	6
Works and Words of our Saviour, gathered from the Four Gospels. By the Rev. H. J. Coleridge, S.J.	0	7	6
Sufferings of the Church in Brittany during the Great Revolution. By Edward Healy Thompson, M.A.	0	6	6
Suppression of the Society of Jesus in the Portuguese Dominions. From Documents hitherto unpublished. By the Rev. Alfred Weld, S.J.	0	7	6
[This volume forms the First Part of the General History of the Suppression of the Society.]			
Teaching and Counsels of St. Francis Xavier. Gathered from his letters. Edited by the Rev. H. J. Coleridge, S.J.	0	5	0
Three Catholic Reformers of the fifteenth Century. By Mary H. Allies.	0	6	0
Thomas of Hereford, Life of St. By Fr. Lestrange	0	6	0
Tribunal of Conscience, The. By Father Gasper Druzbicki, S.J.	0	3	6

RAWES, THE LATE REV. Fr., Edited by.

The Library of the Holy Ghost:—

Vol. I. St. Thomas Aquinas on the Adorable Sacrament of the Altar. With Prayers and Thanksgivings for Holy Communion. Red cloth	0	5	0

Little Books of the Holy Ghost:—

Book 1. St. Thomas Aquinas on the Commandments. 32mo, 233 pp. Cloth gilt	0	2	0
Book 2. Little Handbook of the Archconfraternity of the Holy Ghost. Fourth edition. 111 pp.	0	1	0
Gilt	0	1	2
Book 3. St. Thomas Aquinas on the Lord's Prayer. 139 pp.	0	1	0
Cloth gilt	0	1	3
Book 4. The Holy Ghost the Sanctifier. By Cardinal Manning. 213 pp. 1s. 6d. and	0	2	0
Guide to the Archconfraternity of the Servants of the Holy Ghost. Edited by the Rev. R. Butler, Director, cloth	0	0	2

RICHARDS, REV. WALTER J. B. (D.D.)

Manual of Scripture History. Being an Analysis of the Historical Books of the Old Testament. By the Rev. W. J. B. Richards, D.D., Oblate of St. Charles; Inspector of Schools in the Diocese of Westminster. In Four Parts. each	0	1	0
Or, the Four Parts bound together. Cloth	0	4	0

"Happy indeed will those children and young persons be who acquire in their early days the inestimably precious knowledge which these books impart."—*Tablet*

BURNS & OATES' PUBLICATIONS.

RYDER, REV. H. I. D. (of the Oratory.)
    Catholic Controversy: A Reply to Dr. Littledale's "Plain Reasons." Fifth edition . . . £0 2 6
    "Father Ryder of the Birmingham Oratory, has now furnished in a small volume a masterly reply to this assailant from without. The lighter charms of a brilliant and graceful style are added to the solid merits of this handbook of contemporary controversy."—*Irish Monthly.*

SOULIER, REV. P.
    Life of St. Philip Benizi, of the Order of the Servants of Mary. Crown 8vo 0 8 0
    "A clear and interesting account of the life and labours of this eminent Servant of Mary."—*American Catholic Quarterly.*
    "Very scholar-like, devout and complete."—*Dublin Review.*

THOMPSON, EDWARD HEALY, (M.A.)
    The Life of Jean-Jacques Olier, Founder of the Seminary of St. Sulpice. New and Enlarged Edition Post 8vo, cloth, pp. xxxvi. 628 . . 0 15 0
    "It provides us with just what we most need, a model to look up to and imitate; one whose circumstances and surroundings were sufficiently like our own to admit of an easy and direct application to our own personal duties and daily occupations."—*Dublin Review.*
    The Life and Glories of St. Joseph, Husband of Mary, Foster-Father of Jesus, and Patron of the Universal Church. Grounded on the Dissertations of Canon Antonio Vitali, Father José Moreno, and other writers. Crown 8vo, cloth, pp. xxvi. 488, . . 0 6 0
    "No literature contains a more splendid tribute to St. Joseph."—*Irish Monthly.*
    (By the same author. "Library of Religious Biography." 9 vols: already published. List on application.)

ULLATHORNE, BISHOP.
    Endowments of Man, &c. Popular edition. . . 0 7 0
    Groundwork of the Christian Virtues : do. . . 0 7 0
    Christian Patience, . . do. do. . . 0 7 0
    Ecclesiastical Discourses 0 6 0
    Memoir of Bishop Willson. 0 2 6

WARD, WILFRID.
    The Clothes of Religion. A reply to popular Positivism 0 3 6
    "Very witty and interesting."—*Spectator.*
    "Really models of what such essays should be."—*Church Quarterly Review.*

WISEMAN, CARDINAL.
    Fabiola. A Tale of the Catacombs. . 3s. 6d. and 0 4 0
    Also a new and splendid edition printed on large quarto paper, embellished with thirty-one full-page illustrations, and a coloured portrait of St. Agnes. Handsomely bound. 1 1 0

BURNS AND OATES, LD.
London: Orchard St., W., & 63 Paternoster Row, E.C.;
and at New York.

www.ingramcontent.com/pod-product-compliance
Lightning Source LLC
Chambersburg PA
CBHW051246300426
44114CB00011B/915